SpringerBriefs in Physics

For further volumes:
http://www.springer.com/series/8902

James D. Wells

Effective Theories
in Physics

From Planetary Orbits to
Elementary Particle Masses

 Springer

James D. Wells
PH-TH
CERN
Geneva 23
Switzerland

ISSN 2191-5423   ISSN 2191-5431   (electronic)
ISBN 978-3-642-34891-4   ISBN 978-3-642-34892-1   (eBook)
DOI 10.1007/978-3-642-34892-1
Springer Heidelberg New York Dordrecht London

Library of Congress Control Number: 2012953116

Printed on acid-free paper

Springer is part of Springer Science+Business Media (www.springer.com)

Preface

Effective Theories have been with us since the dawn of science, but it has only been in recent decades that we have found it important enough to give it a clear and voiced name. This new found desire is due in part to our understanding that no finitely written theory is complete. There was a proselytizing impulse among all those who first grasped the vision of Effective Theories. I recall as a Ph.D. student that many fellow students coming out of Boston would repeatedly pepper their conference talks with the words "Effective Theory", and others sometime joked that they were curiously keen on celebrating their ignorance and lamented how sad it was that they had such weak ambition. It was at the time that many particle physicists were proudly espousing their faith in the "Theory of Everything" being around the corner. The extremes of the two camps were in stark contrast.

Today, the culture and language of Effective Theories have permeated all of physics. It is not controversial and not lamentable. The concepts are deeply ingrained in many other areas of theoretical physics. In the subsequent chapters, several different physics subareas are touched upon but the discussions all revolve around Effective Theories. An abstract definition of the term is given in the first chapter, and fleshed out through examples in the following chapters. It is hoped that by the end the reader will have a good feel for how the concepts of Effective Theories affect the thinking of practicing scientists, and can see the power that explicitly agreeing to the Effective Theory mindset can have in developing richer theories of nature and achieving a deeper understanding.

Overview of Subjects Covered

In the following chapters, I wish to emphasize various aspects of Effective Theories across various subdisciplines of physics. Chapter 2 discusses the harmonic oscillator from an Effective Theory point of view. The harmonic oscillator is one of the most important models of physics, and shows up in many guises across all subdisciplines. For this reason I have chosen to start there. The chapter is somewhat allegorical as I go through the story of coming upon a harmonic oscillator system and trying to understand what theory may describe it. The concepts of Effective Theories, and the traps that people may fall into if they do not accept that theories are never complete, are illustrated at each step of the discovery process.

In Chap. 3, I emphasize how blinded we can be to progress if we do not understand that all theories are Effective Theories. I use the example of Newton's law of gravity, and argue that if scientists had the more modern perspective of Effective Theories, they would have not only been quite sure that an anomalous perihelion precession of Mercury would one day be discovered, but they would also have been able to predict roughly what size it would be. As it was the only anomalous precession admitted to the canon after very painstaking experiment and the exhaustion of all other explanations based on mundane effects were analyzed. Reluctantly, the anomaly was accepted and Einstein's theory of gravity ultimately legitimized it.

In no other area of science has Effective Theories played such a prominent role as in elementary particle physics. In Chaps. 4 and 5 I focus on this subfield of science. In Chap. 4, I give a brief introduction to the history of Effective Theories in particle physics before coming to the main theme of Effective Theories and the Higgs boson. The Higgs boson is the elementary scalar particle that is said to give mass to all other known elementary particles. It achieves this by spontaneous symmetry breaking, a concept that will be discussed in some detail. However, the compatibility of Effective Theory ideas and the Higgs boson spontaneous symmetry breaking scale is under dispute. The main purpose of Chap. 4 is to enable the reader to understand what this dispute is and to give various ideas that resolve the dispute. Unlike other chapters, this one contains advanced material that one normally does not encounter until graduate studies. The material is there partly to emphasize to the reader that there is no way to speak intelligibly about the subject without that advanced material. Those who already know the background core material may wish to skip directly to Sects. 4.4 and 4.5 where the focused discussion on the role of Effective Theories is presented.

Finally, in Chap. 5, I show that the concepts of Effective Theory can play an important role in our theory choice activities. The goal of this chapter is to show the culture of theory choice among practicing particle physicists, which is most often not talked about openly among the physicists, and then to describe how the ideas of Effective Theories can change perceptions of what the "Best Theories" are.

The original version of the book was revised: The book has been changed from non-open access to open access and the copyright holder has been updated. The correction to the book is available at https://doi.org/10.1007/978-3-642-34892-1_6.

Acknowledgments

This book was partly born out of lectures to graduate students in Physics and Philosophy of Science, and thus the book is geared mainly for that audience. The material in chapters on the "Harmonic Oscillator" and "Effective Theories in Classical Gravity" was delivered at the University of Michigan in 2010 to a joint Philosophy and Physics graduate seminar. The chapter on "Particle Physics and Effective Theories" is somewhat modified version of lectures delivered to British doctoral students at the University of Cambridge and University of Liverpool in 2008 and 2009. The chapter on "Effective Theories and Theory Choice" was delivered at Philosophy of Science Symposium at the University of Wuppertal in 2012. I would like to thank all of those institutions for their kind hospitality during those visits, and for the excellent discussion I had with students and faculty there. I would also like to thank my University of Michigan and CERN colleagues for the fertile environment in which to discuss these ideas.

Geneva, September 2012   James D. Wells

Contents

Acronyms

ALFB	Adjusted Galileo's Law of Falling Bodies
CERN	Centre Européenne pour la Recherche Nucléaire
ESM	Effectified Standard Model of Particle Physics
GeV	Giga electron volts
GLFB	Galileo's Law of Falling Bodies
IBE	Inference to Best Explanation
km	Kilometers
LHC	Large Hadron Collider
m	Meters
s	Seconds
SM	Standard Model of Particle Physics

Chapter 1
The Utility of Effective Theories

1.1 Definition of Effective Theories and Their Purpose

"Effective Theories" are theories because they are able to organize phenomena under an efficient set of principles, and they are effective because it is not impossibly complex to compute outcomes. The only way a theory can be effective is if it is manifestly incomplete. "Everything affects anything" is generally correct, but it saps confidence in our ability to predict outcomes. Effective Theories modify this depressing maxim by pointing out that "most things are irrelevant for all practical purposes." A tree falling in Peru does not appreciably affect a canon ball's flight in Australia. Any good Effective Theory systematizes what is irrelevant for the purposes at hand. In short, an Effective Theory enables a useful prediction with a finite number of input parameters.

With this definition of Effective Theories it appears that all theories are such, and thus giving it a fancy capitalized name is pointless pedantry. However, the proper name is useful to repeat at times as a reminder that the prominent views of science were not always agreeing that theories were necessarily incomplete, and as a reminder to go beyond it when and if the circumstances may arise. Furthermore, the natural tendency of young students entering science is to believe a theory is either right or useless, when they can never be completely right, but rather merely Effective Theories that are "correct enough for our purposes in this domain." Frequent and formalized reminders of this are helpful for newcomers to the field.

The other purpose of emphasizing the name Effective Theories is to force us to confront a theory's flaws, its incompleteness, and its domain of applicability as an integral part of the theory enterprise. The most useful Effective Theories are ones where we know well their domains of applicability, and can parametrically assess the uncertainties induced by ignoring the "irrelevant." They may even have a well-defined procedure for becoming more and more complex as one wishes to compute to higher accuracies. This is the case in many Effective Field Theories of particle physics, such as pion scattering or even graviton scattering. There is

J. D. Wells, *Effective Theories in Physics*, SpringerBriefs in Physics,
DOI: 10.1007/978-3-642-34892-1_1, © The Author(s) 2012, corrected publication 2022

a science in understanding the circumstances of when questions can be addressed using accurate, convenient Effective Theories, and it is generally acknowledged that scale separation (Hillerbrand 2013) is one important feature of systems that enable an Effective Theory to separate out well the "relevant" from the "irrelevant". Indeed the phrase "irrelevant operator" is a technical term used in particle physics (Cohen 1993) to identify small contributions to phenomena caused by dynamics at a much different energy scale than is being probed. This issue arises in one form or another in all Effective Theories and will be seen in the examples presented.

1.2 Galileo's Law of Falling Bodies as an Effective Theory

Throughout this book we will get progressively more modern in our discussion of how to apply the concepts of Effective Theories to physics. We will move from the harmonic oscillator to Newton to Einstein to Fermi to Higgs and others. Before we do that, let us begin in this introductory chapter with Galileo—one of the first scientists who had what is recognizable as a modern perspective to scientific thought. Galileo was dedicated to knowing what was correct with less care about his or others' preconceived ideas. He was dedicated to experimental verification as an unbiased arbiter of theories. He investigated many things, but we will focus on his theory of falling bodies, and within that context show, as a warm-up to more sophisticated theories later, how the concepts of Effective Theory could have engendered further insight into a more general theory of gravity beyond just describing a falling body.

Let us suppose that we are back in the day of Galileo, well before Newton came along, and we are very mathematically sophisticated for the times. Upon reading Galileo's book the *Two Sciences* we come across the following passage:

> When, therefore, I observe a stone initially at rest falling from an elevated position and continually acquiring new increments of speed, why should I not believe that such increases take place in a manner which is exceedingly simple and rather obvious to everybody? If now we examine the matter carefully we find no addition or increment more simple than that which repeats itself always in the same manner. This we readily understand when we consider the intimate relationship between time and motion; for just as uniformity of motion is defined by and conceived through equal times and equal spaces (thus we call a motion uniform when equal distances are traversed during equal time-intervals), so also we may, in a similar manner, through equal time-intervals, conceive additions of speed as taking place without complication; thus we may picture to our mind a motion as uniformly and continuously accelerated when, during any equal intervals of time whatever, equal increments of speed are given to it.... And thus, it seems, we shall not be far wrong if we put the increment of speed as proportional to the increment of time; hence the definition of motion which we are about to discuss may be stated as follows: A motion is said to be uniformly accelerated, when starting from rest, it acquires, during equal time-intervals, equal increments of speed (Galileo 1638).

In mathematical language Galileo is saying $\delta v = g\delta t$, where v is the speed and g is the constant of proportionality. In differential calculus language $\delta v, \delta t \rightarrow dv, dt$.

Bringing dt to the other side of the equation one can rewrite Galileo's Law as $dv/dt = g$. But change in velocity with respect to time is nothing other than the acceleration, and Galileo's law becomes $a = g$, which is "uniform acceleration" as Galileo himself called it. Notice that the mass of the stone falling is not in this equation. More on that later. Another way to write the above equation is

$$\ddot{z} = -g \text{ (Galileo's Law of Falling Bodies),} \qquad (1.1)$$

in the convention that z is the position of the ball with increasing z in the opposite direction of the acceleration vector.

As an aside, every first year physics student has computed the trajectory of a ball in a uniform gravitational field. The equation of motion is usually derived from Newton's Second Law of Motion $F = ma$. In this case the force is $-mg$ where $g = 9.8\,\text{m/s}$ is the acceleration downward due to gravity on the Earth's surface, and $a = \ddot{z}$ is the second time derivative of the ball's motion—the actual acceleration of its trajectory. The equation of motion is then $\ddot{z} = -g$, which is exactly Galileo's Law. Despite everyone knowing this, the reader is here requested to forget the more sophisticated later era of Newton, where this particular equation $\ddot{z} = -g$ is a simple derivation of a deeper law. Instead, I would like to ask the reader to treat $\ddot{z} = -g$ as a law of nature that has no parent—it is something stand-alone discovered by Galileo. That is why I am giving it a fancy name: "Galileo's Law of Falling Bodies", or GLFB for short. Let us press forward with GLFB, and ask what Effective Theories may say about it.

To give us something concrete to talk about with regard to GLFB, let us compute the time it takes for a body at rest to drop from a height h. The position of the body as a function of time is

$$z(t) = h - \frac{1}{2}gt^2. \qquad (1.2)$$

Falling a distance h then takes time $T = \sqrt{2h/g}$. Notice, this does not depend on the mass of the body—an interesting conclusion that Galileo understood well. He knew that air friction caused bodies to slow down, and he even understood the concept of terminal velocity,[1] but most impressively he realized that air friction was a complication that was not fundamental to the problem:

> Now seeing how great is the resistance which the air offers to the slight momentum [*momento*] of the bladder and how small that which it offers to the large weight [*peso*] of the lead, I am convinced that, if the medium were entirely removed, the advantage received by the bladder would be so great and that coming to the lead so small that their speeds would be equalized (Galileo 1638).

In other words, in the limit that the density of the body was much higher than the density of the air, the air friction was not important. Galileo repeated this principle

---

[1] "... there is no sphere so large ... or so dense... that the resistance of the medium, although very slight, would check its acceleration and would, in time reduce its motion to uniformity" (Galileo 1638).

in other places, and understood it well: the fundamental law of falling bodies with resistance-less medium is uniform acceleration.

Another demonstration of Galileo's genius was that he understood better than anyone at that time that scientific claims were not only about deep thoughts that sounded good, but required experiment to test them and that any result was subject to question. At one point he took a swipe at Aristotle for holding what Galileo thought was an unjustified opinion: "... I greatly doubt that Aristotle ever tested by experiment whether it be true ..." (Galileo 1638). Galileo was certainly no respecter of persons, but rather had unswerving loyalty to determining what was correct. Even when he introduced his theory of falling bodies he qualified it by saying, "we shall not be far wrong" if we agree to his theory. Tentativeness, testing and refinement, the hallmarks of science, were important to his approach.

Galileo surely would not have minded any correction to his law that was not in conflict with what appeared to be sacrosanct symmetries of nature, such as invariance under rotations and space and time translations (Arnold 1989). A correction that seems quite reasonable is to disrupt uniform acceleration slightly by adding a correction term that depends on height position z.[2] Thus let us add the correction $\ddot{z} = -g + cz$, where c is some "small" constant.

The constant c is unknown and so this theory is not very predictive. However, we can make some intelligent guesses of roughly what value it could take. For one, we know that somehow we have to make cz have units of acceleration. This requires c to have units of acceleration/length. This is an awkward set of units. However we can simplify it by utilizing the one and only constant of our original theory, which is g and has units of acceleration. Thus, the obvious thing to do is let $c \rightarrow g/R$, where R is some unknown fixed constant of length. What could R possibly be? The test bodies are being pulled to earth, and they are all being pulled with (nearly) uniform acceleration independent of the size of the test body,[3] and so it is very reasonably to assume that we need to look to the Earth to provide us with a "natural length scale" to assign R. The radius of the Earth, $R_e = 6400$ km, is the obvious candidate.[4]

If we were dogmatic and very arrogant we would say that our choices were "obvious" and that this new law, the Adjusted GLFB (ALFB), is the correct first correction and write $\ddot{z} = -g(1 - z/R_e)$ and then start computing. However, let us be humble scientists and suggest that this correction is perhaps "not far wrong", as Galileo might say, and insert a "constant of tentativeness" η, which is dimensionless

---

[2] This is not in conflict with Galilean translation invariance, as z is shorthand for a *difference* in position of the body with respect to the earth's surface $z = r - R_{earth}$.

[3] Furthermore, using the size of a small test body as the parameter R would lead to dramatically too large effects, and for that reason also it can be dismissed as an option.

[4] There are several other length scales that perhaps might be equally justified, including the circumference of the earth ($R = 40,000$ km), the height of the tallest mountain ($R = 9$ km), or the depth of the deepest sea ($R = 11$ km). The latter two are perhaps less intuitively relevant and could be dismissed as serious candidates. Nevertheless, if one kept an open mind to them all, the length scales are all within about a factor of 10^3 of each other, which might appear disastrously large to estimate a correction term, but it is decidedly better than not knowing how to estimate within a factor of ∞.

and perhaps not far from 1. Our new ALFB can be written as

$$\ddot{z} = -g\left(1 - \eta\frac{z}{R_e} + \cdots\right) \quad \text{(Adjusted Galileo's Law of Falling Bodies)}. \quad (1.3)$$

Writing theories down with extra terms that have "natural sizes" and are consistent with symmetries is a cornerstone of the Effective Theory approach. This example is intended to demonstrate that a new theory can be generated by having this mindset, and the new theory is more correct, even if a little less predictive.

Ignoring the higher order "$\cdots$" terms, the solution to the problem of position as a function of time now becomes

$$z(t) = h\cosh\left(\sqrt{\eta\frac{g}{R_e}}\,t\right) - \frac{1}{2}gt^2 \quad (1.4)$$

and the time it takes to reach $z = 0$ is

$$T = \sqrt{\frac{2h}{g}}\left(1 + \frac{\eta}{2}\frac{h}{R_e} + \mathcal{O}\left(\frac{h^2}{R_e^2}\right)\right). \quad (1.5)$$

A body dropped from $200\,\text{m}$ takes about a tenth of a second longer according to ALFB with $\eta = 1$ compared to the $6.5\,\text{s}$ predicted by the GLFB.

In an alternative scientific history this effect of longer dropping time could have been measured and the anomaly noted before Newton's theory of gravity was decisively understood. The measurements would have converged on $\eta = 2$ to within experimental uncertainties. A discrepancy with Galileo's pure GLFB would not have been the subject of deep worries about human's ability to understand the laws of the universe since Galileo himself was tentative about his law. In time, Newton's theory would then develop, and the value of η would be computed to be exactly 2, and Newton's law of gravity would then replace GLFB as the overarching theoretical framework by which to understand and compute the trajectories of falling bodies.

We have seen from this simple example that one does not need to know the more fundamental theory of Newtonian gravity to anticipate corrections, compute their effects, and compare with data. The Effective Theory of ALFB is better than Galileo's original law, despite being less predictive, because ultimately it can accommodate the data better and reflects Newton's deeper theory. We will see another example of this in the chain of theories in a later chapter that shows how one could have anticipated phenomenological implications of Einstein's General Relativity by taking a more tentative, Effective Theory approach to Newton's Law of Gravity.

References

Arnold, V.I.: Mathematical Methods of Classical Mechanics. Springer, New York (1989) (For an excellent discussion of the principles of "Galilean Invariance")

Cohen, A.G.: Selected topics in effective field theories for particle physics. In: Proceedings of TASI 1993: The Building Blocks of Creation, Boulder (1993)

Galileo, G.: Dialogues Concerning Two New Sciences 1638. Translated by de Salvio A., Crew, H. Macmillan, New York (1914), Web edition published by eBooks[at]Adelaide

Hillerbrand, R.: Explanation via microreduction. On the role of scale separation in quantitative modeling. In: Falkenburg, B., Morrison, M. (eds.) Why More is Different. Philosophical Issues in Condensed Matter Physics and Complex Systems. Springer, Heidelberg (2013)

Chapter 2
Harmonic Oscillator as an Effective Theory

Abstract The concepts of Effective Theories are illustrated allegorically within the context of one of the most ubiquitous models of oscillating physical phenomena—the harmonic oscillator.

2.1 Basics of the Harmonic Oscillator

The concepts and issues related to effective theories can be illustrated quite nicely by the harmonic oscillator problem. The harmonic oscillator is one of the most ubiquitous mathematical models of physics phenomena. It is present in almost every system with a restoring force, which includes the galaxy, solar system, springs, atoms, molecules, and innumerable other configurations.

The main point I would like to illustrate is that the lowest order effective potential for the harmonic oscillator is an excellent approximation to the motion of a system over a wide range of amplitudes. However, at some point it breaks down when the amplitude is large enough, and then control over the system is lost unless a deeper theory is understood. We shall not go into the construction of deeper theories in this chapter, but rather focus on the domain of applicability of the harmonic oscillator effective theory, and show how small corrections can be anticipated and then measured by precise experiments to start building a more complete picture of the potential governing the system.

To keep the illustration simple, we will restrict ourselves to one-dimensional harmonic motion of a particle subject to the restoring potential $V(x) = kx^2/2$. The Lagrangian of the system is then

$$L = \int dt \left(m\frac{\dot{x}^2}{2} - k\frac{x^2}{2} \right). \tag{2.1}$$

J. D. Wells, *Effective Theories in Physics*, SpringerBriefs in Physics,
DOI: 10.1007/978-3-642-34892-1_2, © The Author(s) 2012, corrected publication 2022

From the principle of least action the equation of motion gives Newton's second law of motion $F = ma$ the form

$$m\ddot{x} = -kx \implies m\ddot{x} + kx = 0. \tag{2.2}$$

Defining $\omega^2 = k/m$, we can rewrite this as

$$\ddot{x} + \omega^2 x = 0 \tag{2.3}$$

which has the solution

$$x(t) = A\sin(\omega t) \tag{2.4}$$

where A is the amplitude, and the boundary condition of $x = 0$ at $t = 0$ is enforced.

Let us review a few basic facts about the harmonic oscillator solution. The period is

$$T_{period} = \frac{2\pi}{\omega} = 2\pi\sqrt{\frac{m}{k}}. \tag{2.5}$$

The amplitude A of motion is related to the initial velocity by equating full potential energy at maximum amplitude to the full kinetic energy at maximum velocity:

$$\frac{1}{2}mv_{max}^2 = \frac{1}{2}kA^2 \implies A = v_{max}\sqrt{\frac{m}{k}} = \frac{v_{max}}{\omega} = \frac{v_{max}T_{period}}{2\pi}. \tag{2.6}$$

It should also be noted that the period of the harmonic motion is not dependent on the amplitude of the motion. This is clear from Eq. 2.5 where it is shown that the period only depends on the input parameters m and k. The amplitude and maximum velocity conspire with each other such that v_{max}/A is always equal to $\sqrt{k/m}$.

2.2 Ubiquity of the Harmonic Oscillator

The harmonic oscillator problem is ubiquitous in physics, describing small motions of an object attached to a string, molecules vibrating in crystals, electrical circuit response, etc. There is a straightforward reason why there are so many examples that follow simple harmonic behavior. Let us suppose that the equilibrium point (i.e., the minimum of the potential) is about the origin. Then, the potential for motion is a power series of the form

$$V(x) = a_2x^2 + a_3x^3 + a_4x^4 + \cdots . \tag{2.7}$$

We do not write down a constant term or a term linear in x because the first is irrelevant and the second term cannot be present if $x = 0$ is a local minimum. If it is present, one shifts x to cancel it, which is the place of the new extremum.[1] There are an infinite number of potentials that can be written down, with various relative weightings of x^4, x^{12}, etc. The motions of a particle or entity about the equilibrium can be very different depending on the potential.

Nevertheless, the universal quality of harmonic motion is ubiquitous because at values of x below some critical value x_{crit} the potential is always dominated by the x^2 term. For example, in comparing the $a_2 x^2$ term to the $a_3 x^3$ term, the ratio is

$$\frac{a_2 x^2}{a_3 x^3} = \frac{a_2}{a_3} \frac{1}{x} \implies a_2 x^2 \text{ term dominates over } a_3 x^3 \text{ when } x < x_{crit} = \frac{a_2}{a_3}. \quad (2.8)$$

In other words, small enough amplitudes are always very well described by simple harmonic motion in a x^2 potential.

In the following we will investigate an abstract system that has harmonic oscillation in the "low-energy limit", when the amplitude is small. We shall see that through a combination of precision measurements and venturing into the high-energy unknown we can learn more about the system. In the course of these investigations I wish to give a sense of the usefulness of thinking in terms of effective theories, as well as seeing the limitations of it.

2.3 First Theory

Let us suppose that there exists a System[2] that appears to be undergoing harmonic oscillation. For simplicity, the System will be chosen to have lengths of amplitude and times for the period of motion to be measured most conveniently in meters and seconds; however, this is only for intuitive concreteness, and one can multiply these units by orders of magnitude in any direction as appropriate for different systems.

In the earliest stages of investigation of the System we see that it is undergoing oscillatory behavior with a period of about 10 s. The resolution of the instrumentation is not good enough to resolve any deviations from pure harmonic motion, and so we posit that the motion is governed by the potential

$$V(x) = \frac{x^2}{2} \implies \ddot{x} + \omega^2 x = 0 \quad \text{(Theory 1)}. \quad (2.9)$$

Let us now suppose that we try to test this theory by precision measurements. Again, at the early stages of experimenting on a system, the resolution may not be so good. Let us suppose that is the case for our simple System, and assume that the period is measured to be

---

[1] If for some reason $a_2 = 0$, then a_3 will need to be zero also, otherwise $x = 0$ is not a local minimum, and the first term to worry about is x^4. This is a complication that we need not worry about for now.

[2] We capitalize System to give it a reference name for rest of the discussion.

$$T_{period} = 10\,\text{s} \pm 0.3\,\text{s} \quad \text{(Measurement 1)}. \tag{2.10}$$

This period of motion can be accommodated by our theory as long as

$$\omega = 0.63 \pm 0.02\,\text{s}^{-2} \quad \text{(Parameter Fit 1)}. \tag{2.11}$$

It is no mystery that we could find a value of ω that fit the period. No matter how well we measure the period, it is only one observable and the theory has one parameter that can always be adjusted to match it. We need more observables to test the validity of the theory more fully.

2.4 Second Theory

Another drawback of having just one observable is that there are an infinite number of theories that we could write down trivially whose parameters could be adjusted in an infinite continuum of values to accommodate the measurement. One such theory has the same potential as Theory 1 except for now we add an x^3 correction term to the potential,

$$V(x) = k\frac{x^2}{2}\left(1 + \frac{2x}{3\Lambda_A}\right) \implies \ddot{x} + \omega_A^2 x\left(1 + \frac{x}{\Lambda_A}\right) = 0 \quad \text{(Theory 2)} \tag{2.12}$$

where ω_A and Λ_A, a new length scale, are two parameters that can have a relation between them that give the same period. Here are two values:

$$\omega_A = 0.63\,\text{s}^{-2} \quad \text{and} \quad \Lambda_A = \infty \tag{2.13}$$

$$\omega_A = 0.631\,\text{s}^{-2} \quad \text{and} \quad \Lambda_A = 250\,\text{m} \quad \text{(Parameter Fit 2)} \tag{2.14}$$

where the first line is equivalent to Theory 1 and the second line is just one parameter fit out of an infinite number of possibilities.

Upon close inspection of Theory 2 we notice that the correction term always generates a force of the same direction no matter what the value of x: it pushes the particle away from the origin when x is negative and pulls it back to the origin when $x > 0$, whereas the first term always is restoring. This should create an asymmetry in the time it takes for the Particle to cross $x = 0$ half-way through its full periodic motion compared to the time it takes to cross $x = 0$ again on its second half of the motion. We can compute this difference in time. Even though the total period $T_{period} = 10\,\text{s}$ stays the same, the first and second halves of the distance covered by the motion would be asymmetric if x/Λ_A is not too suppressed:

$$T_{period}^{+1/2} \neq T_{period}^{-1/2} \quad \text{but} \quad T_{period} = T_{period}^{+1/2} + T_{period}^{-1/2} = 10\,\text{s}. \tag{2.15}$$

Therefore, an important additional observable to measure are these "half periods" to see if they are antisymmetric as Theory 2 predicts.

Let us now suppose that there are improvements in the experimental instrumentation such that we can measure each "half period", $T_{period}^{+1/2}$ and $T_{period}^{-1/2}$, and it can be done to accuracies of 0.01 s. And let us suppose that after some time of measurement it is determined that

$$T_{period}^{+1/2} = 5.05\,\text{s}, \quad T_{period}^{-1/2} = 5.06\,\text{s}, \quad \text{and}$$

$$T_{period} = 10.11 \pm 0.01\,\text{s} \quad \text{(Measurements 2)}. \tag{2.16}$$

To within the error bars of 0.01 s the two period halves are equal.

The usual scientific approach to the present situation is to say that the simpler model wins out if it accommodates the data as well as the more complicated theory. Thus, the community of scholars faced with the measurements above may well conclude that Theory 1 is correct, or conclude that even if the x/Λ_A term is present it is so suppressed that it is immaterial to the physics.

As we shall discuss later, this is the kind of statement that one might find in particle physics when considering higher dimensional operators of Standard Model particles. As in particle physics we may hold firm to the idea that there is no reason why these extra terms should not exist. Indeed, in an effective theory the full series expansion of additional terms should exist. But we must acknowledge that their coefficients may be too small to discern from our experiments.

2.5 Fancy Explanations

Not seeing the effects of the asymmetric x/Λ_A term after greatly improving the experimental situation to look for it would likely get the community thinking hard for the reasons of that failure. As we already mentioned, the diehard believers would just say that Λ_A has a value just higher than the experimental sensitivities would see. Others would invent reasons for why x/Λ_A should never have been there in the first place. These reasons need to be based on some kind of symmetry argument.

There are two straight-forward symmetry arguments that would banish the x/Λ_A correction to the potential. The first argument is to presume that the potential has an $x \to -x$ discrete symmetry. This would banish all odd corrections that could give rise to asymmetric half periods. Our next correction would then be x^2/Λ^2. We will investigate the experimental consequences of that potential shortly.

Another symmetry argument that says the harmonic oscillator lagrangian is exact with a conformal symmetry, $x \to \lambda x$ where λ is some arbitrary scaling parameter. Although the Lagrangian is not invariant under this, the equations of motion are. It is this scaling symmetry that tells us that time observables are independent of the spatial scaling. In other words, the (time) period is independent of the (spatial) amplitude.

There is a temptation of smart people to promote the most sophisticated and fancy arguments to explain the phenomena. It is not very sophisticated to say "the additional terms are too small to see". But it is fancy to say things like "conformal symmetry" and "discrete symmetry." And if the experimental situation languishes long enough theorists can become even more sophisticated with their description of why these terms must be banished, and look down upon people who do not catch the fever of fancy explanations. And if it goes on even longer it will be so entrenched in the highest schools of the land, that few will want to challenge it by proposing ways to find evidence for non-fancy corrections to the spatial scale-invariant theory.

2.6 Third Theory

Nevertheless, let us suppose that we take courage and wish to press forward in testing Theory 1 yet again. Odd corrections may exist, but we may need orders of magnitude more precision to see evidence for $T_{period}^{+1/2} \neq T_{period}^{-1/2}$. We may have more luck introducing only even power corrections to the potential. So we shall do this by introducing

$$ V(x) = k\frac{x^2}{2}\left(1 - \frac{x^2}{2\Lambda_B^2}\right) \implies \ddot{x} + \omega_B^2 x\left(1 - \frac{x^2}{\Lambda_B^2}\right) = 0 \quad \text{(Theory 3)} \quad (2.17) $$

What can we do to test and try to strain the theory? We know that measuring the half-periods does no good. However, being excellent students of the prevailing scale-invariant idea, we know that the period should not change depending on the amplitude. We need to find a way to perturb the system to increase the amplitude and see if the period changes.[3]

Let us suppose in our system that the particle passes through the origin with velocity of 10 m/s. Changing it requires significant technical skill, but we find a way to do it. We increase the energy into the system and obtain a new initial velocity of 15 m/s, which increases the amplitude by approximately 50%. Upon measuring the period we get

$$ T_{period} = 10.25\,\text{s} \pm 0.01\,\text{s} \quad \text{(Measurement 3)} \quad (2.18) $$

which differs by many standard deviations from the 10.11 s value obtained when $v_{initial} = 10$ m/s, and is a clear signal for breaking of the spatial scale invariance of

---

[3] It is here I would like to remind the reader again that this is a fanciful allegory to how experiment and theory interplay on the effective theory stage, and although a simple macroscopic harmonic motion system can be manipulated and measured in all sorts of ways with ease, sometimes other systems are significantly more challenging to do the analogy of measuring half periods or of increasing the amplitudes.

the equations of motion. This is the first firm proof that the exact harmonic motion law of $V(x) \propto x^2$ is not fully respected.

We are likely to be quite excited about this, because we posited a theory that said there should be violations of scale invariance when the amplitude grows. And now that we see it we want to fit the parameters. Here is one such choice that works well

$$\omega_B = 0.63\,\text{s}^{-2} \quad \text{and} \quad \Lambda_B = 95\,\text{m} \quad \text{(Parameter Fit 3).} \tag{2.19}$$

The two measurements at two different velocities are accommodated by these two choices of parameters.

Theory 3 is "better" than the old simple harmonic oscillator law of Theory 1, because it accounts for all the data. It accounts for equal half periods, and accounts for the measurements when the initial velocity is at $v = 10\,\text{m/s}$ and at $v = 15\,\text{m/s}$. However, Theory 3 is not the only theory that could do this. We could have had an x^6 correction, for example, that would have fit just as well this limited amount of data. Dissatisfaction may set in that we cannot be confident of any precise formulation of the theory to describe the system. If arbitrary corrections are allowed now, then anything goes.

This is both the beauty and the frustration of effective theories. Being committed to the notion that all terms should be allowed in a potential consistent with the symmetries we believe to be sacrosanct, and then test them with ever increasing experimental sophistication, has given us insight that deviations from the pure harmonic oscillator potential are possible. However, these ideas of effective theory appear to have muddied the waters rather than have led to "the theory." We come to the realization that this is one of the limitations of effective theories. By itself it cannot raise you to a deeper physical insight. It is merely a statement that all operators (i.e., all corrections) should be added to your theory and then experiment can measure or put limitations on the couplings. However, if you do happen onto a deeper theoretical insight, that can put order to all the operators that may arise.

2.7 Deep Theory Conjecture

Now let us suppose that we let our success get to our heads, and we become supremely confident that we know of a deeper theory to explain the data. Nevermind how we came to it—that is not important here—but suppose the deep theory we become convinced of is

$$V = \omega_T^2 L_T \left[1 - \cos(x/L_T)\right] \implies m\ddot{x} + \omega_T^2 L_T \sin(x/L_T) = 0 \quad \text{(Theory 4).} \tag{2.20}$$

The data that has been taken to date suggests that

$$\omega_T = 0.63\,\text{s}^{-2} \quad \text{and} \quad L_T = 38.8\,\text{m} \quad \text{(Parameter Fit 4).} \tag{2.21}$$

We note that there is no difference between Theory 3 predictions and Theory 4 predictions as long as the initial speed stays below 20 m/s and the timing resolution is not better than 0.01 s.

However, we can make a bold prediction based on our new deep and fundamental theory conjecture: if the initial velocity is doubled to 30 m/s the period jumps to 11.23 s, whereas for Theory 3 the prediction is 11.36 s. Experimentalists may puzzle over how to double the initial velocity for many years, but finally are able to do it. When they collect the data, they find $T_{period} = 11.35$ s ± 0.01 s, which is a dramatic confirmation of Theory 3, and the hubris of the conjecturing Theory 4 is defeated.

2.8 Ultimate Test?

After the extreme test of Theory 3, which was years in the making and passed so decisively and impressively, the smart people figure out lots of fancy language to explain why it had to be true and what symmetry properties it has. It is written in every textbook. However, there was one more experiment that people wished to do. For years it has been suggested that if you are able to reach initial speeds greater than 42 m/s the Particle will never come back. In other words, the initial energy will be so great that it will exceed the confining potential barrier of Theory 3. However, getting to 42 m/s is a technological nightmare, and it will take decades to do it.

But let us suppose that after decades of R&D, it has been figured out how to launch the particle to speeds of 50 m/s from $x = 0$. When the experiment is conducted the particle flies off into the unknown. Twenty seconds go by, one minute goes by, an hour goes by, days and months go by, and the particle has never returned. Scientists are not surprised, but a little disappointed. It would be so much fun for a new anomaly to happen, but the theory looks solid and inviolate.

The scientists may move on, and study other things like sandpiles and solar flares. But one day, many years later, the particle returns! And nobody knows why, except a bright young student who realizes that the next term in the effective potential may have been what returned it.

Chapter 3
Effective Theories of Classical Gravity

Abstract If the concepts underlying Effective Theory were appreciated from the earliest days of Newtonian gravity, Le Verrier's announcement in 1845 of the anomalous perihelion precession of Mercury would have been no surprise. Furthermore, the size of the effect could have been anticipated through "naturalness" arguments well before the definitive computation in General Relativity. Thus, we have an illustration of how Effective Theory concepts can guide us in extending our knowledge to "new physics", and not just in how to reduce larger theories to restricted (e.g., lower energy) domains.

3.1 Introduction

The purpose of these lectures is to introduce the concepts of Effective Theories to students of Philosophy, Mathematics and Physics who have a shared interest in the philosophy and history of physics. The concept I wish to discuss, Effective Theory, is a thoroughly modern notion; nevertheless, I wish to illustrate it with a very old and intuitively accessible problem in physics: Mercury's anomalous perihelion precession.

Le Verrier announced in 1845 a small discrepancy in the precession rate of Mercury's perihelion compared to Newton's theory, even after taking into account all the disturbing influences throughout the solar system such as the effect of other planets' orbits.[1] This came as a surprise, and more or less nobody believed at the time that it was the fault of Newton, but rather the fault of observers who had not seen the other celestial bodies that must surely be perturbing Mercury's orbit. Historically, that is the beginning of the problem. Le Verrier believed that an as-yet unobserved mass distribution inside the orbit of Mercury was the source

---

---

[1] In 1859 Le Verrier gave a number for this advance: 35 arcseconds per century (Le Verrier 1859). It was later reevaluated by S. Newcomb (Newcomb 1882), who determined the correct value of 43 arcseconds per century.

J. D. Wells, *Effective Theories in Physics*, SpringerBriefs in Physics,
DOI: 10.1007/978-3-642-34892-1_3, © The Author(s) 2012, corrected publication 2022

of the discrepancy. He and others advocated the existence, for example, of a new small planet ("Vulcan" as it was sometimes called) that would be observed when astronomers developed the instruments necessary to find it (Roseveare 1982). Such was not the case. By the 1890's it became clear to most that new large-scale object(s) was not the explanation (Oppenheim 1920), despite some ill-fated protestations otherwise (Poor 1921). The resolution of the problem came with Einstein's General Relativity, which predicted precisely the 43" of arc per century observed, and the case was closed.

However, I want to argue that anticipation of the "problem" could have occurred much before Le Verrier. What prevented scientists from anticipating Mercury's perihelion precession was not lack of mathematical skill, or lack of experimental abilities. It was solely due to not having the right mindset. Unlike perhaps in decades and centuries gone by, no competent scientist should retain an unfailing commitment to any theory. All theories are incomplete, even given that some theories are better than others. The code phrase of this mindset is Effective Theories. The concept is a powerful one that has born much fruit in theories of particle physics, condensed matter systems, and even cosmology.

These notes are meant to be a somewhat pedagogical and technical exposition of the Mercury problem and the application of Effective Theory ideas to the problem. In some parts of this lecture I will follow an "alternative history" path with the scientists Alice and Bob who vaguely understand the importance of Effective Theories and who will devise a theory that can accomodate the perihelion precession rate well before Einstein's General Relativity comes along, and may even be able to predict roughly the numerical rate of the precision and make predictions for other planets through "naturalness" arguments. The latter could have been possible after diligent reflections on the philosophical challenges of Newton's theory. I will compute the General Relativity rate at the end, in order to show how elegantly it comes out of that more complete theory, and to show that it matches the Effective Theory "predictions" by Bob and Alice. And finally I will conclude with some more remarks on the meaning of the results.

3.2 Orbits in Newton's Theory

To remind some students who have not seen celestial mechanics for some time, we begin with the computation of particle orbits in Newton's gravity. The reader familiar with these basics should feel free to skim the section only for definitions and conventions that I will use later.

We know that the orbits predicted by Newton's law of gravity are respected quite well by the planets, and so any change in the equations of motion for the orbits will need to be small perturbations. In Newton's gravity, a test particle with mass m orbits around a particle of mass $M \gg m$ according to the equations of motion derived from the lagrangian

$$L = \frac{1}{2}m\dot{\mathbf{r}}^2 + \frac{\alpha}{r} \qquad (3.1)$$

where $\alpha = GMm$ with G being Newton's constant. M represents the sun's mass in this lecture, and m the planet's mass, most often Mercury. It is appropriate to assume that $M \gg m$ such that any correction is negligible due to the difference of m from the reduced mass $\mu = Mm/(M + m)$, which is technically the precise mass one should use in the kinetic energy term. Lagrange's equations of motion are

$$m\ddot{\mathbf{r}} = -\frac{\alpha}{r^2}\hat{\mathbf{r}} \qquad (3.2)$$

where $\hat{\mathbf{r}}$ is the unit vector in the $\mathbf{r}$ direction.

3.2.1 Orbital Solution

The lagrangian is rotationally invariant, and so the motion of the particle is most conveniently evaluated by casting the vector equation of motion into the two polar component equations

$$m(\ddot{r} - r\dot{\phi}^2) = -\frac{\alpha}{r^2} \quad \text{(radial equation)} \qquad (3.3)$$

$$m(2\dot{r}\dot{\phi} + r\ddot{\phi}) = 0 \quad \text{(angular equation)}. \qquad (3.4)$$

The second equation is equivalent to

$$\frac{d}{dt}(mr^2\dot{\phi}) = 0 \qquad (3.5)$$

which implies that $mr^2\dot{\phi}$ is a constant in time. At the apogee (furthest) or perigee (closest) point of the orbit the radius vector $\hat{\mathbf{r}}$ is exactly perpendicular to the angular vector $\hat{\phi}$ and the magnitude of the angular momentum vector $\ell = \mathbf{r} \times \mathbf{p}$, where $\mathbf{p} = mr\dot{\phi}\hat{\phi}$, is exactly $mr^2\dot{\phi}$. Since angular momentum is conserved and $mr^2\dot{\phi}$ is conserved, if they are equal at one point they are equal at all points in the orbit. Thus, the constant value inside the time derivative Eq. (3.5) is none other than angular momentum: $\ell = mr^2\dot{\phi}$. This also proves that the motion is in a plane. Since angular momentum is a conserved vector quantity, the direction must also be preserved which is only possible if $\mathbf{p}$ perpetually lies in the same plane as $\mathbf{r}$. This justifies our evaluation of a three-dimensional problem in terms of just two variables (r, ϕ) in the plane of motion.

Let us now solve the radial differential equation to obtain an exact solution of the orbit for particle m. By rewriting $r \equiv 1/u$, recasting all time derivatives as $d/dt \to \dot{\phi} d/d\phi$ when possible, and recognizing that $\dot{\phi} = l/r^2 m$ from conservation of angular momentum, one finds that the governing differential equation of motion is

$$\frac{d^2 u}{d\phi^2} + u = \frac{\alpha m}{\ell^2}. \tag{3.6}$$

Interestingly, this equation takes the form of the harmonic oscillator equation. The solution is

$$u(\phi) = u_0 \cos \phi + \frac{\alpha m}{\ell^2}, \tag{3.7}$$

where u_0 is a constant that is not determined by the theory but the particular circumstances (i.e., initial conditions) of the system. In terms of the more direct variable r, the solution is

$$r(\phi) = \frac{\rho}{1 + e \cos \phi}, \quad \text{where } \rho = \frac{\ell^2}{\alpha m}, \quad \text{and } e = u_0 \rho, \tag{3.8}$$

and it is assumed that $\phi = 0$ is at perigee. ρ is sometimes called the lactus rectum of the orbit.

The constant e is called the eccentricity with which one can classify an orbit as circular ($e = 0$), elliptical ($0 < e < 1$), parabolic ($e = 1$), or hyperbolic ($e > 1$). Focusing on the $0 \le e < 1$ case of elliptical or circular orbits, we find that

$$r_{\min} = \frac{\rho}{1 + e}, \quad \text{and } r_{\max} = \frac{\rho}{1 - e}. \tag{3.9}$$

The relation between the semimajor axis a of the elliptical orbit and the other variables is given by

$$a = \frac{r_{\min} + r_{\max}}{2} \quad \text{which implies} \quad \rho = a(1 - e^2). \tag{3.10}$$

3.2.2 The Hamiltonian and V_{eff} Description

An alternative way to approach the problem is to compute the Hamiltonian and consider the orbit from the perspective of a one-dimensional effective potential for radial motion. I provide the very basics of this to remind the students of the formalism which is used by some papers relevant to the perihelion precession. We first expand the lagrangian in terms of radial and angular coordinates starting from the identity

$$\dot{\mathbf{r}}^2 = \dot{r}^2 + r^2 \sin^2\theta \dot{\phi}^2 + r^2 \dot{\theta}^2$$
$$\Longrightarrow \dot{r}^2 + r^2\dot{\phi}^2 \text{ (valid in the } \sin\theta = 1 \text{fixed orbital plane)} \quad (3.11)$$

The Hamiltonian is constructed as

$$H = \sum_i \dot{q}_i p_i - L \quad (3.12)$$

using the momentum factors

$$p_r = \frac{\partial L}{\partial \dot{r}} = m\dot{r}, \text{ and } p_\phi = \frac{\partial L}{\partial \dot{\phi}} = mr^2\dot{\phi}, \quad (3.13)$$

which implies

$$H = \frac{p_r^2}{2m} + \frac{p_\phi^2}{2mr^2} - \frac{\alpha}{r}. \quad (3.14)$$

The Hamiltonian is independent of ϕ, which implies from Hamilton's equations of motion,

$$\dot{p} = -\frac{\partial H}{\partial q}, \text{ and } \dot{q} = \frac{\partial H}{\partial p}, \quad (3.15)$$

that $\dot{p}_\phi$ is a conserved quantity:

$$\dot{p}_\phi = \frac{\partial H}{\partial \phi} = 0 \Longrightarrow p_\phi = \text{const.} \quad (3.16)$$

This of course is just a restatement of the conservation of angular momentum

$$\ell = p_\phi = mr^2\dot{\phi}. \quad (3.17)$$

We can substitute Eq. 3.17 back into Eq. 3.14, which gives a one-dimensional Hamiltonian as promised:

$$H = \frac{p_r^2}{2m} + \frac{\ell^2}{2mr^2} - \frac{\alpha}{r}. \quad (3.18)$$

The Hamiltonian is a constant of the motion—the energy of the system—and it is useful sometimes to consider the dynamics of particle motion from this considera-tion where $E \equiv H = T + V$ and T and V are the kinetic and potential energies respectively. Here the potential for the one dimensional motion is often called the effective potential and is given by

$$V_{eff}(r) = \frac{\ell^2}{2mr^2} - \frac{\alpha}{r}. \tag{3.19}$$

It is this potential that governs the radial potential with the first term pushing the particle away from the origin and the second term attracting the particle to the origin. The balance giving orbital motion between two turning points of zero radial kinetic energy, the apogee and perigee.

3.3 Perihelion Precessions from Perturbations

From the previous section we know that the orbit from Newton's simple $1/r^2$ force law is

$$u(\phi) = \frac{1}{r(\phi)} = \frac{1}{\rho}(1 + e \cos \phi). \tag{3.20}$$

This obviously does not allow any advancement of the perihelion. The minimum is where $du/d\phi = 0$, which implies $\sin \phi = 0$ and therefore $\phi = 0, 2\pi, 4\pi, \ldots$ mark the successive perihelions. The discovery of the anomalous perihelion precession of Mercury, if it can be established, would signal the end of the Newtonian era and initiate the search for a better theory. As the reader is no doubt aware, perihelion precessions exist for every planet's orbit (see Table 3.1), but for the present let us continue on our theoretical discussion.

3.3.1 $1/r^2$ Correction to the Central Potential

Let us look at how the orbits change if we add a $1/r^2$ correction to the potential of the gravitational interaction lagrangian. Let us call this Bob's theory with lagrangian

$$L = \frac{1}{2}m\dot{\mathbf{r}}^2 + \frac{\alpha}{r}\left(1 + \frac{R_{bob}}{r}\right) \tag{3.21}$$

where $\alpha = GMm$, with G being Newton's constant, M is the mass of the sun, and m is the mass of the planet under consideration. This new law requires the introduction of a new fundamental length scale R_{bob}, which is a priori unknown. However, we do know, as will be shown below, that it leads to a perihelion precession of the orbits governed by this law.

Lagrange's equation of motion for this theory is

$$\text{radial}: m(\ddot{r} - r\dot{\phi}^2) = -\frac{\alpha}{r^2}\left(1 + \frac{2R_{\text{bob}}}{r}\right) \tag{3.22}$$

$$\text{angular}: m(2\dot{r}\dot{\phi} + r\ddot{\phi}) = 0 \tag{3.23}$$

The angular equation yields conservation of angular moment $\ell = mr^2\dot{\phi} = \text{const}$ just as before. Using this, we can write the radial differential equation as

$$\frac{d^2u}{d\phi^2} + u = \frac{\alpha m}{\ell^2}(1 + 2R_{\text{bob}}u) \tag{3.24}$$

This can be rewritten as

$$\frac{d^2u}{d\phi^2} + \left(1 - \frac{2R_{\text{bob}}}{\rho}\right)u = \frac{1}{\rho}, \quad \text{where } \rho = \frac{\ell^2}{\alpha m}. \tag{3.25}$$

The general solution to this equation, assuming perihelion is placed at $\phi = 0$, is

$$u(\phi) = u_0 \cos\left(\phi\sqrt{1 - \frac{2R_{\text{bob}}}{\rho}}\right) + \frac{1}{\rho - 2R_{\text{bob}}}, \tag{3.26}$$

or, written differently,

$$u(\phi) = \left(\frac{1}{\rho - 2R_{\text{bob}}}\right)\left[e\cos\left(\phi\sqrt{1 - \frac{2R_{\text{bob}}}{\rho}}\right) + 1\right] \tag{3.27}$$

where $e = u_0(\rho - 2R)$.

The $u(\phi)$ solution describes the motion of a precessing ellipse. The first perihelion by definition is at $\phi = 0$ and the second perihelion occurs when

$$\phi\sqrt{1 - \frac{2R_{\text{bob}}}{\rho}} = 2\pi \implies \phi = \frac{2\pi}{\sqrt{1 - \frac{2R_{\text{bob}}}{\rho}}} = 2\pi + 2\pi\frac{R_{\text{bob}}}{\rho} \tag{3.28}$$

The small perihelion advance is the deviation of ϕ from 2π and is $\delta = 2\pi R_{\text{bob}}/\rho$.

Given our previous computations, we are now able to evaluate the relationship between the extra length scale R_{bob} and the perihelion advance of an orbit. In one case, if we have made a measurement of the perihelion advance, we can derive what value R_{bob} must be to reproduce that value

$$R_{\text{bob}} = (1.16\,\text{km})\left(\frac{\delta/T_{\text{orbit}}}{\text{arcsec} \cdot \text{century}^{-1}}\right)\left(\frac{\rho}{1\,\text{au}}\right)\left(\frac{T_{\text{orbit}}}{1\,\text{year}}\right) \tag{3.29}$$

where ρ is related to the common parameters of the semimajor axis $a = (r_{\text{min}} + r_{\text{max}})/2$ and eccentricity e by the relation $\rho = a(1 - e^2)$.

Table 3.1 Data for planetary orbits

Planet	T_{orbit} (years)	e	a (au)	ρ (au)	r_{min} (au)	r_{max} (au)
Mercury	0.241	0.206	0.387	0.371	0.307	0.467
Venus	0.615	0.007	0.723	0.723	0.718	0.728
Earth	1.000	0.017	1.000	1.000	0.983	1.017
Mars	1.881	0.093	1.524	1.511	1.382	1.666
Jupiter	11.86	0.048	5.203	5.191	4.953	5.453
Saturn	29.46	0.056	9.539	9.509	9.005	10.07
Uranus	84.02	0.047	19.19	19.15	18.29	20.09
Neptune	164.8	0.009	30.06	30.06	29.79	30.33
Pluto	247.7	0.249	39.46	37.01	29.63	49.29

T_{orbit} is the time for one full revolution in earth years, e is the eccentricity of the orbit, a is the semimajor axis in astronomical units (1 au $= 1.496 \times 10^{11}$ m), $\rho = \ell^2/GMm^2 = a(1 - e^2)$ is the orbital latus rectum in astronomical units (and is independent of m ultimately), $r_{min} = a(1 - e)$ is the distance of perigee in astronomical units, and $r_{max} = a(1 + e)$ is the distance of apogee in astronomical units

On the other hand, if we have a theory for what R_{bob} should be, we can make a prediction for the perihelion advance in units of arc seconds per century:

$$\frac{\delta}{T_{orbit}} = \frac{2\pi R_{bob}}{\rho T_{orbit}} = (0.866\, \text{arcsec} \cdot \text{century}^{-1}) \left(\frac{1\,\text{au}}{\rho}\right) \left(\frac{\text{years}}{T_{orbit}}\right) \left(\frac{R_{bob}}{1\,\text{km}}\right) \quad (3.30)$$

3.3.2 $1/r^3$ Correction to the Central Potential

Alice's theory has a $1/r^3$ correction to the potential

$$L_{alice} = \frac{1}{2}m\dot{\mathbf{r}}^2 + \frac{\alpha}{r}\left(1 + \frac{R_{alice}^2}{r^2}\right), \quad (3.31)$$

which gives a $1/r^4$ correction to the gravitational force law. Lagrange's equations for her theory are

$$\text{radial}: m(\ddot{r} - r\dot{\phi}^2) = -\frac{\alpha}{r^2}\left(1 + \frac{3R_{alice}^2}{r^2}\right) \quad (3.32)$$

$$\text{angular}: m(2\dot{r}\dot{\phi} + r\ddot{\phi}) = 0 \quad (3.33)$$

Here again the angular momentum $\ell = mr^2\dot{\phi}$ is conserved from the angular equation, and the radial equation becomes

$$\frac{d^2u}{d\phi^2} + u = \frac{\alpha m}{\ell^2}(1 + 3R_{\text{alice}}^2 u^2). \tag{3.34}$$

We'll solve this equation employing techniques of perturbation theory. We treat the last term of Eq. 3.34 as a small perturbation and solve first the equation

$$\frac{d^2u}{d\phi^2} + u = \frac{\alpha m}{\ell^2} \tag{3.35}$$

which is just the standard Newtonian orbit solution

$$u_N(\phi) = \frac{1}{\rho}(1 + e\cos\phi), \quad \text{where } e = u_0\rho \tag{3.36}$$

where the subscript N refers to the Newtonian solution, u_0 is an initial condition constant and $\rho = \ell^2/\alpha m$ is the usual value.

The next step is to now substitute $u \to u_N + \delta u$ into Eq. 3.34 where we only keep one order in perturbation theory. Since u_N part of this expression cancels the usual part of the differential equation from Newton's law, we are left with a differential equation for the perturbation δu:

$$\frac{d^2\delta u}{d\phi^2} + \delta u = \frac{3}{\rho}R_{\text{alice}}^2 u_N^2(\phi) \tag{3.37}$$

$$= \frac{3}{\rho^3}R_{\text{alice}}^2(1 + 2e\cos\phi + e^2\cos^2\phi) \tag{3.38}$$

To obtain the complete solution we need to solve for δu. The theory of ordinary differential equations tells us that all we need is *any* particular solution, and here is one:

$$\delta u = \frac{3}{\rho^3}R_{\text{alice}}^2\left(1 + e\phi\sin\phi + \frac{e^2}{3}\cos 2\phi + e^2\sin^2\phi\right) \tag{3.39}$$

The perihelians of the orbit can be obtained by solving for ϕ in

$$\rho\frac{du}{d\phi} = -e\sin\phi + 3\frac{R_{\text{alice}}^2}{\rho^2}$$
$$\times\left(e\sin\phi + e\phi\cos\phi - \frac{2}{3}e^2\sin 2\phi + 2e^2\sin\phi\cos\phi\right) = 0 \tag{3.40}$$

The existence of the $\phi\cos\phi$ term in this equation, which came from the $\phi\sin\phi$ term in δu, is causing the perihelion on the next cycle to shift away from 2π. Defining $\phi = 2\pi + \delta$ we can solve for δ in the perturbative expansion:

$$\delta = 6\pi\frac{R_{\text{alice}}^2}{\rho^2}. \tag{3.41}$$

Given our previous computations, we are now able to evaluate the relationship between the extra length scale R_{alice} and the perihelion advance of an orbit. In one case, if we have made a measurement of the perihelion advance, we can derive what value R_{alice} must be to reproduce that value

$$R_{\text{alice}} = (7.58 \times 10^6 \, \text{m}) \left(\frac{\rho}{\text{au}} \right) \left(\frac{T_{\text{orbit}}}{\text{years}} \right)^{1/2} \left(\frac{\delta / T_{\text{orbit}}}{\text{arcsec/century}} \right)^{1/2} \qquad (3.42)$$

On the other hand, if we have a theory for what R_{alice} should be, we can make a prediction for the perihelion advance in units of arc seconds per century:

$$\frac{\delta}{T} = \frac{6\pi R_{\text{alice}}^2}{\rho^2 T_{\text{orbit}}} = (1.74 \, \text{arcsec} \cdot \text{century}^{-1}) \left(\frac{R_{\text{alice}}}{10^7 \, \text{m}} \right)^2 \left(\frac{1 \, \text{au}}{\rho} \right)^2 \left(\frac{1 \, \text{yr}}{T_{\text{orbit}}} \right). \qquad (3.43)$$

3.4 Philosophical Challenges to Newton's Theory

We pause here to describe some foundational questions that Newton's theory faced. There are three main philosophical problems: (1) What is the nature of absolute time and space, and is it necessary to invoke it? (2) Why should the gravitational mass be equal to the inertial mass in the equations of motion? And (3) how does nature enable action at a distance responses?

Regarding Absolute Space and Time, Newton sets forth his ideas in the first Scholium of *Principia*. Almost immediately upon the publication of his book, Newton faced criticism from noted physicists and mathematicians. The most famous adversary regarding this was Leibnitz, who claimed that the only thing that need be talked about, and which ultimately defines space and time, is the relative motions of objects (relativism). Appeals to absolutes make no sense. Newton's friend Samuel Clarke argued vociferously for the absolute viewpoint (substantivalism). Their correspondences are famous, and illuminating in the history of science. Over time these discussions progressed from what some might think is word quibbling to important physics principles emphasized by Mach and Einstein to name just two. Pedantic rigor of thinking can lead to the thought processes that generate significantly better theories, and this philosophical problem is arguably an illustration of that.

The second problem, why is the gravitational mass equal to the inertial mass in my mind is the problem that should have kept everyone sleepless for those many centuries when there was not an answer. Newton's theory has nothing to say on the matter, except *well, there it is*. These masses are two separate beasts, and why they should be the same? The resolution of this issue is one of the core motivating principles behind General Relativity, which succeeds in giving a deeper explanation for this curious equality.

The third philosophical problem is sometimes called the problem of action at a distance. There are two aspects of action at a distance. The first is why should two bodies far removed from each other with nothing in between them feel gravitational attraction. Should not there be some "touching" or medium that carries the gravity force from one body to the other? This action at a distance occurs between particles separated by a large vacuum of nothing. This is hard to take. Even Newton was disturbed by it, especially the latter aspect. In 1693 he wrote his friend Richard Bentley "It is inconceivable that inanimate brute matter should, without the mediation of something else which is not material, operate upon and affect other matter without mutual contact ...." (Thayer 1953).

The second aspect of the problem, which is related to the first, is how can two bodies far removed from each other in space instantaneously feel the effect of another's gravitational force. Newton's theory implicitly assumes that all particles feel all other particles' gravitational attraction strength by the exact separations of those particles at each moment of time. If a particle moves just a little, everybody knows about it instantly and the resolution of forces are adjusted instantly. To Newton and others, action at a distance was intolerable, but the Newtonian system was the best thing going, and it had tremendous practical value, so it was not to be abandoned despite its flaws.

The issue of instantaneity was noted from the start, and Laplace touched upon it in his highly influential *Traité de Mécanique Céleste*, published from 1797 to 1825. He stated that instantaneous propagation did not appear convincing,[2] and noted that Bernoulli had suspicions as well. But Laplace knew that if the propagation were indeed finite it would have to be extraordinarily fast, and even suggested, incorrectly as it turns out, that some observations imply that it is eight million times that of light. Laplace briefly brought up the possibility of modifying the inverse square law based on this potential objection but ultimately dismissed it, stating that the simplicity of Newton's theory authorizes us to think of it as a rigorous law of nature.[3]

Nevertheless, the philosophical challenges to Newton's theory are enough to realize that it was not a complete theory. As we say often in physics today, there must be "physics beyond the Standard Model". How might signal of "new physics" show up beyond Newton's theory? Let us consider, for example, the disturbing underlying assumption of action at a distance. As we implied above, there are two different issues with action at a distance. There is the aspect of reaching across the mediumless vacuum, and there is the aspect of instantaneous transmission of information to all particles in the universe when one particle moves.

Transforming our theory from reaching across the vacuum action at a distance to action by local contact is the subject of the theory of fields. Particles source fields that permeate spacetime, and other particles experience those fields. Thus, action at a distance is replaced by particle-field interactions in this classical point of

---

[2] "La propagation instantanée qu'ils supposaient à cette force me parut peu vraisemblable" (Laplace 1805).

[3] "En général, on verra dans le cours de cet ouvrage que la loi de la gravitation réciproque au carré des distances représente avec une extrême précision toutes les inégalités observées des mouvement célestes: cet accord, joint à la simplicité de cette loi, nous autorise à penser qu'elle est rigoureusement celle de la nature" (Laplace 1805).

view. The emanating field propagates at finite velocity, which is incorporated self-consistently into modern field theories, retaining causality and introducing the more acceptable action by local contact.

We do not need to fast forward all the way to the field theories of today to ask how Newton's theory can be pressured experimentally by applying our philosophical worries. The most obvious way one should have thought to do it is by testing the instantaneous aspect of action at a distance. If one doubts that it is to be rigorously upheld, then we should expect that a quick movement of a body in a mechanical system might yield unexpected results since it might be significantly displaced from its original position by the time the other bodies "get word" of its flight, and it becomes ambiguous to know what direction and magnitude of force should be applied at all times. Thus, at some sufficiently high speed we might expect to see something unusual—something unplanned for in the Newtonian world. The trouble is, we do not know a priori what speed this breakdown would occur, and we certainly do not know what new description would be applicable.

In circumstances like this, it is often best to write down effective theories that satisfy the symmetries of your worldview and do precision measurements to find deviations. The pattern of deviations or the values of couplings in the effective theory can lead to new insight when explained by a deeper theory. Bob's $1/r^2$ correction theory and Alice's $1/r^3$ correction theory to the gravity potential in the preceding sections do precisely that. They are Galilean invariant, and satisfy all the symmetries cherished even then: rotational invariance and translation invariance.

We apply this approach of writing down corrections to planetary motion because this is our greatest hope to find cracks in the old classical world view. Since no cherished symmetries are violated by the additional terms we have found before, we may even expect to find breakdowns of Newton's theory by the orbits of the planets, especially since they are accessible and moving faster with respect to each other and the sun than any laboratory system that could have possible been created on the earth at the time. Precision measurements of fast planetary motions thus had good reason to be the first place to find breakdown of Newton's theory. No planet moves faster than Mercury. Indeed, it is Mercury where the first fissures arise, as we shall describe in the following sections.

3.5 Effective Theories

It is my contention that the concepts of Effective Theories, if understood and held by the early Newtonian scientists, would have led to a prediction that there must *necessarily* be an anomalous perihelion precession of Mercury and other planets, and that even the order of magnitude could have been guessed well before Le Verrier's announcement in 1859. There was no barrier to adopting these ideas in Newton's day, as it requires no new special experimental knowledge, nor knowledge of Einstein's relativity, but rather a more mature approach to how we think about the laws of nature. In order to present this viewpoint, I shall first give a précis of the modern notions of Effective Theories.

At its core, the term Effective Theory is short for a body of evidence that has led us to understand that "everything depends on everything else" may be true in principle but certainly not true in practice. In a restricted domain, the theory manifests symmetries and properties that provide the ability to calculate observables without the requirement of making reference to features outside the domain. A simple example of this is that we can compute the trajectory of a football to any practical precision without needing to know the location of Uranus. The effects of Uranus on the trajectory are suppressed by a relative factor of $\frac{m_e r_e^2}{m_U d_U^2} \sim 3 \times 10^{-14}$, where r_e is the radius of the earth, d_U is the distance from Uranus, and m_e (m_U) is the mass of the earth (Uranus). This is much too small to take into account for any practical need. The diminishing effect of Uranus as $d_U \to \infty$ is the principle of decoupling, which is at the core of Effective Theory utility and is the central reason why science works and we are able to compute and predict observables.

A central concept of Effective Theory is the recognition that a full theory with heavy and light degrees of freedom can be written at low energies in terms of just light degrees of freedom after "integrating out" the heavy states or "coarse graining" over the small scales. We use "heavy" and "light" abstractly here, as it could refer to masses, momenta, velocity, etc. The chiral lagrangian of QCD, the Fermi theory of electroweak interactions, the Landau-Ginzburg theory of superconductivity (Polchinski 1992) can all be recognized as an Effective Theory of a more fundamental theory.

This top-down approach to understanding Effective Theories can give us a multitude of theoretical insights into the nature of simplified low-energy theories. It is this top-down approach that is traditionally how the power of Effective Theory concepts is demonstrated in particle physics (Cohen 1993; Rothstein 2003), fluid mechanics (Delgado-Bucalioni et al. 2005), material science (Abrams 2005), and essentially any other field that has a separation of scales. However, when considering theories from bottom up, the concepts we learn from Effective Theories can help us deduce modifications and additions to our present theories that can be tested by experiment. Success then can lead to motivations for inducing a more fundamental theory that reproduces the Effective Theory when restricted to its domain. It is this direction in theory analysis that I emphasize here for our present purposes.

The insight that I would like to focus on, which I believe is the most powerful one when it comes to divining additions and modifications to theories, is the role that symmetries and naturalness play in the construction of the "complete" Effective Theory. A symmetry is a recognition that something (a triangle, an equation, etc.) stays the same even if you make a closed set of transformations (i.e., group operations) on that object (rotations by 180°, interchange of x and y variables, etc.). All of our fundamental theories have inherent recognized symmetries in them. We cannot proceed without these recognitions in the Effective Theory, because even the names we give to objects are merely shorthand notation for their symmetry properties (e.g., electrons are spin-1/2 representations of the Lorentz Group with additional gauge symmetry representation labels).

One of the principle consequences of the Effective Theory approach to establishing natural law is that all possible interactions (or "terms") consistent with the recognized symmetries of the Effective Theory are generically expected. There may or may not be additional terms that violate the symmetries, but terms that do not violate the symmetries must be included. In the realm of Effective Theories within quantum field theory, Weinberg, reflecting on the last three decades of work on the subject, has made the equivalent point that an Effective Theory may be considered self-consistent and not sick "as long as every term allowed by symmetries is included" (Weinberg 2009).

In short, the precise form of a theory or law is not what is to be taken most seriously—it is the recognized symmetries. Upon sorting out the symmetries, the Effective Theory is to be developed with all possible terms consistent with the symmetry, and then qualitative expectations for experiment can be presented. What remains is measurement and pinning down the actual values of the coefficients to each symmetry preserving interaction term.

3.5.1 Application to Newton's Gravitation

Newton's law of gravitation is that the force between two bodies of masses m and M is inversely proportional to the square of the distance between them, with the proportionality constant being Newton's constant G:

$$F(r) = \frac{GMm}{r^2}, \quad \text{or} \quad V(r) = \frac{GMm}{r} \tag{3.44}$$

where $V(r)$ is the potential. In Book 3 of Principia, Newton states categorically that the inverse square law is "proved with the greatest exactness from the fact that the aphelia are at rest" and that "the slightest departure from the ratio of the square would necessarily result in a noticeable motion of the apsides...." (Newton 1999). Thus, the theory was created and solidified as a proposition to the world.

Newton's inverse-square law was so sacrosanct that few would ever doubt it. Immanuel Kant in 1747 used the inviability of the inverse-square law to derive that space had three dimensions. This is due to what we would say today is the conservation of gravitational flux lines emanating from a point mass through the surface of a sphere of arbitrary radius. God could have chosen a different gravity law, Kant says, and the number of spatial dimensions then would have had to be different.[4]

---

[4] "Zweitens, dass das Ganze, was daher entspringt, vermöge dieses Gesetzes [inverse-square law] die Eigenschaft der dreifachen Dimension habe; drittens, dass dieses Gesetz willkürlich sei, und da Gott dafür ein anderes, zum Exempel des umgekehrten dreifachen Verhältnisses [i.e., inverse-cube law], hätte wählen können; dass endlich viertens aus einem andern Gesetze auch eine Ausdehnung von andern Eigenschaften und Absmessungen geflossen wäre" (Sect. §10 in Kant 1747).

This rigid adherence to "god-given" specific law is ultimately incorrect reasoning, and it is in conflict with modern views of Effective Theories.

The modern sensibility says that we should focus more on the symmetries, and then refashion the complete Effective Theory using them. What are the symmetries of the Newtonian world? The symmetries are that the laws of physics cannot be affected by one's orientation in space, by one's location in space, nor by one's location in time. The laws must be invariant to any transformation of rotation, spatial translation, or time translation. These symmetry properties go under the name of Galilean invariance. As a side comment, the Lorentz invariance of Einstein's special relativity asymptotes to Galilean invariance in the low velocity limit (i.e., when $v \ll c$).

The interaction term of Eq. 3.44 is merely one term in an infinite number of terms that could be written down that are completely consistent with Galilean invariance. An Effective Theory approach would be to introduce them all and investigate the consequences. There is no meaningful symmetry that demands only the inverse square law interaction. Assured of this, one example would be to embellish Newton's law by

$$V_{ET}(r) = \frac{GMm}{r} \left[1 + \sum_{n=1}^{\infty} \lambda_n \left(\frac{r_0}{r} \right)^n \right] + \cdots \qquad (3.45)$$

where r_0 is some dimensionful Effective Theory length scale and λ_n are dimensionless coefficients, which together with r_0 can be found by performing precise experiments. We should note that there are an infinite variety of other terms that could be added, including r^j and $\dot{r}^k$ interactions, but we streamline the argument by looking only at one class of corrections that decouple as $r \to \infty$.

3.5.2 Inevitable Perihelion Precession

An extremely important conclusion can already be presented from the rules of Effective Theories. Any deviation from the pure inverse square law will lead to a perihelion precession of the planets, and as the constructed Effective Theory demands additions to the inverse square law there will be an anomalous perihelion precession of the planets. On the other hand, we know that the inverse square law is approximately correct and thus we have added terms that decouple as $r \gg r_0$. The perihelion precession of Mercury is very small, and so we expect that r_0 should be much less than the orbital radius of Mercury around the sun. In that case, we are justified in looking at the first-order corrected potential, which we can write as ($\lambda_1 r_0 \to R$):

$$V_1(r) = \frac{GMm}{r} \left(1 + \frac{R}{r} \right). \qquad (3.46)$$

By these arguments of Effective Theory, an anomalous perihelion precession of Mercury is inevitable. It is only a question of what value does R take, which then sets the numerical value of the precession. In the subsequent sections we discuss some arguments for what R might be, from the vantage point of pre-special relativity and pre-general relativity days, and make rough quantitative predictions for the precession rate.

Up to this point we have argued that the focus should have been more on the symmetries of the gravitational theory rather than the concretization of the theory. A more complete Effective Theory for Newtonian gravity would have been accepted and one would have fully expected anomalous perihelion precessions of the planets. A potential similar in form to Eq. 3.46 would have been put forward, and the task of theoretically divining or experimentally measuring R would have been the consuming activity.

3.6 Mercury's Anomalous Perihelion Precession

Let us imagine that Bob and Alice are two physicists who are working in the post Le Verrier and pre Einstein era. They are smitten by the Newtonian worldview. They do not wish to do radical things to explain this perihelion precession. They are well-versed in the concepts of Galilean Invariance, Hamilton's Principle, and have an inkling of the ideas of effective theories. Naturally, they want to describe this precession through a Galilean invariant effective theory of gravity. Bob announces that he wishes to add a $1/r^2$ correction to the lagrangian. Not wanted to follow in Bob's footsteps, Alice declares that the force law should be even powers of $1/r^2$ and so her first correction to the lagrangian is $1/r^3$. The two lagrangians are

$$L_{\text{bob}} = \frac{1}{2}m\dot{\mathbf{r}}^2 + \frac{\alpha}{r}\left(1 + \frac{R_{\text{bob}}}{r}\right) \tag{3.47}$$

$$L_{\text{alice}} = \frac{1}{2}m\dot{\mathbf{r}}^2 + \frac{\alpha}{r}\left(1 + \frac{R_{\text{alice}}^2}{r^2}\right) \tag{3.48}$$

where $\alpha = GMm$, with G being Newton's constant, M is the mass of the sun, and m is the mass of the planet under consideration. These are the two lagrangians of Bob and Alice that we studied in a previous lecture. These new laws of Bob and Alice require the introduction of a new fundamental length scale R_i. They do not know what that length scale is, but they have hopes that the new data will pin it down for them.

Before we look more closely at Bob and Alice's theories, we should remark again that in the classical history of gravity, there were early attempts to explain anomalies by changing Newton's laws, even in the manner of Alice and Bob. Such theories go under the name of "Clairaut laws". Clairaut proposed in 1745 that Newton's law should be corrected by a $1/r^4$ force term in order to explain some thought-to-be anomalies in the movement of the lunar perigee. However, he found in the end there was not a discrepancy, which buried such laws deeper into the dustbin of history.

Table 3.2 Anomalous perihelion precession rates of the planets compared to expectations from Newton's law of gravity and taking into account all other sources of precession (effects of other planets orbits, etc.) (Duncombe 1956)

Planet	δ/T (arcsec/century)
Mercury	43.11 ± 0.45
Venus	8.4 ± 4.8
Earth	5.0 ± 1.2

More modern references test gravity (including precession rates) through the parameters of the so-called parametrized post-Newtonian (PPN) approach (Will 2005)

Newcomb commented in 1882 that such laws were "out of the question" because they disrupted the gravitational strength so wildly at very close distances where the correction term would come to dominate (Newcomb 1882). As late as 1910 Newcomb, the world's leader on this issue, was stating that all the data up to that point "... seems to preclude the possibility of any deviation from that law [Newton's inverse-square law]" and that Mercury's perihelion advance is best explained by "the hypothesis of Seeliger" (Newcomb 1910), which was a zodiacal light theory that contained intra-Mercurial distributions of orbital matter minimally disruptive to all other astronomical observations except Mercury's perihelion advance (see, e.g., Chap. 4 of (Roseveare 1982)).

Bob and Alice's theory are a return to the Clairaut law in some ways. In the next few subsections we merely state the effect they would have on planetary orbits. After a discussion of Effective Theories and how they apply to this problem, we shall proceed with a somewhat fanciful alternative history of how deviations from Newton's laws could have been explained and interpreted from the point of view of Effective Theories after the anomaly was announced by Le Verrier. But it should be kept in mind, and will be emphasized again in the concluding section, that these theories could have been anticipated, and perhaps even should have been anticipated, before Le Verrier's announcement.

3.6.1 Analyzing Bob's $1/r^2$ Correction Theory

From Eq. 3.30 we can compute in Bob's theory that it is necessary that $R_{bob} = 4.4$ km if Mercury is to have the measured 43 s of arc per century in its perihelion precession. Given this value of R_{bob}, Bob can make predictions for the perihelion advance of other planets. Using Eq. 3.28 he finds $\delta/T_{orbit} = 8.6''$ of arc per century for Venus's perihelion precession and $3.8''$ for the earth. These predicted values compare favorably to the measurements for Venus and Earth presented in Table 3.2. The predictions are well within the errors, and Bob is pleased because he has found a way to explain the anomaly while yet retaining Galilean invariance as a fundamental symmetry of spacetime. He has done this through the means of a simple expansion correction to Newton's law of gravity. Nothing radical was done.

Despite the successes, Bob is not totally satisfied. He wants to know if he can argue for this new length constant in nature R_{bob}. It's a very strange distance 4.4 km. He wonders how he can formulate this distance from all the invariants swirling around him. It should not depend on the mass of each planet, he reasons, because we have just shown that one value of R_{bob} appears to work universally well for all planets. The other options we have to build a length scale are from Newton's constant G, the mass of the Sun M and angular momentum. Bob fails to find any natural combination that will give 4.4 km.

Before giving up he recalls that his intuition has told him that there is some characteristic high speed such that Newton's simple laws become strained (see Sect. 3.4). He does not know what that speed value is, and his new law is just as much action at a distance as the old one, but he carries on by giving this new speed a name, v_{bob}. With this new undetermined speed in hand he realizes immediately that he can form a new length scale GM/v_{bob}^2. Can this be the origin of R_{bob}? What value must v_{bob} be to recover $R_{bob} = 4.4$ km? A simple calculation yields

$$v_{bob} = \sqrt{\frac{GM}{R_{bob}}} = 1.7 \times 10^8 \text{ m/s}. \tag{3.49}$$

This quantity v_{bob} that Bob has derived is a very curious number! His colleagues down the hall have been working on the theory of electromagnetic phenomenon and a speed very close to that keeps showing up in their equations, $c = 3.0 \times 10^8$ m/s. This is the propagation speed of light. He decides this cannot be a coincidence, but he is not sure what to make of it. He decides to define a new scale based on these thoughts, the "sun's electro-gravity scale" $R_{EG} \equiv GM/c^2$. R_{bob} can now be written in terms of this definite scale $R_{bob} = \lambda_{bob} R_{EG}$. It is very curious that the data fits very well if $\lambda_{bob} = 3$ is an integer. He writes on a piece of paper his new theory of gravity

$$L_{bob} = \frac{1}{2} m \dot{\mathbf{r}}^2 + \frac{GMm}{r} \left(1 + 3 \frac{GM/c^2}{r} \right), \tag{3.50}$$

and he is pleased with its simplicity, elegance and symmetry. He does not know how the speed of light c crept in, but he is satisfied since his lagrangian looks "natural" given that there are no really big or really small numbers populating it. Furthermore, he knows that if he must construct a new length scale with a speed, the "natural" next known threshold of speed is the speed of light, and so this correction is "natural" to explore. He feels he is on to something big.

Bob finds another interesting connection with this scale GM/c^2. He recognizes that there is a small radius R_E of a infinitesimal (i.e., radius less than R_E) spherical body of mass M for which an object going the speed of light would not be able to escape. This light-speed trapping radius is a curiosity: if light were corpuscular in any sense, as Newton and others thought it might be, then we could see no light emanating from within the radius R_E of the massive body. This sets a mystery scale to gravity that requires further scrutiny and may be a length scale associated with changes in gravity. The computation of this scale is simple in the Newtonian world, and is

$$R_E = \frac{2GM}{c^2} \quad \text{(light non – escape radius)} \tag{3.51}$$

This is only a factor of two different than the value of R he has derived from the perihelion precession rate. It should be noted that R_E is the precisely the Schwarzschild radius derived in General Relativity, which is a well-known special scale for spherically symmetric objects for more reasons than just what was stated above (Schwarzschild 1916; Wald 1984). Furthermore, it should be recalled that the speed of light was being quantitatively estimated (Rømer 1676) even before Newton's *Principia*, and by 1729 it was known to within a few percent (Bradley 1729), and so this scale had precise meaning from the very beginning days of Newtonian gravity.

Despite these interesting connections, Bob gets nervous looking over his equations. Equation 3.27 seems to indicate that if $\rho < 2R_{bob} = 6GM/c^2$, the orbits do not make sense anymore, as the equations formally say $r < 0$ which is nonsensical. He relaxes briefly when he realizes that $2R_{bob}$ is only 9 km, which is well below the orbital radius of any planet, and furthermore it is even below the radius of the sun, which is 7×10^5 km. Thus, there is no danger that some small object rotating around the sun would have no chance to be described by Bob's theory, since it would be inside the sun.

Nevertheless, he is still a bit uncomfortable. Nowhere in his derivation was the radius of the sun ever required. In principle, all that mass of the sun could have been at one infinitesimal point for all the equations knew. Nevermind how to pack all that mass in with a radius less than 9 km, it is a possibility in principle that such a tightly packed object exists, and if it did, there is no way his theory could describe close-by orbits with characteristic orbital latus rectum size $\rho < 9$ km. He knows his theory cannot be the end all of all the theories anyway due to not knowing why c crept into his equations, despite that being the natural next "speed scale" to consider, but now he is even more discomfited because he can imagine configurations where his theory just cannot even give an answer. But that is for another day. He has succeeding in explaining the precessions of Mercury, Venus and Earth and that is enough for a day's work. And that is what Effective Theories do. They explain the day's work—Bob clearly has made progress—but there is more to be learned and understood. Effective Theory practitioners understand that all possible questions cannot be resolved instantly, and that there are necessarily deeper effective theories to come.

3.6.2 Analyzing Alice's $1/r^3$ Correction Theory

Alice now wishes to make definite her lagrangian with $1/r^3$ potential corrections by specifying the value of R_{alice} from Mercury's anomalous perihelion precession and then predicting what the other precession rates are. Upon fitting Mercury data she finds $R_{\text{alice}} = 9.04 \times 10^5$ m. Using that fixed value for all planets she then predicts $\delta/T_{\text{orbit}} = 4.43''$ of arc per century for Venus and $1.4''$ of arc per century for the Earth. The Venus result is nearly 2σ off compared to the measurement, and the Earth result is about 3σ off of the measurement (see Table 3.2). Alice has a choice now. She can say her theory predicts that further refined measurements of the precession rates will yield smaller central values of the precession rates for Venus and Earth in concert with her theory. Or, she can take the 3σ discrepancy seriously and attempt to modify her theory.

Alice makes the right choice and seeks to modify the theory. She computes what R_{alice} needs to be for each planetary case to precisely hit the measured values. She finds

$$R_{\text{alice}}^{\text{mercury}} = 90 \times 10^7 \text{ m}, \quad R_{\text{alice}}^{\text{venus}} = 1.3 \times 10^7 \text{ m}, \quad R_{\text{alice}}^{\text{earth}} = 1.5 \times 10^7 \text{ m}. \quad (3.52)$$

Similar to Bob, she begins to think about how these length scales can be identified with all the quantities that she has available to her in the problem: $M, m_{\text{planet}},$ and ℓ. She cannot come to a satisfactory answer. These constants alone are not enough to form the length scales of Eq. 3.52.

However, in Alice's trials she notices something interesting. The R_{alice} lengths are proportional to angular momentum divided by mass of the planet, $R_{\text{alice}}^i \propto \ell_i/m_i$, with the same proportionality constant. This constant has the dimensions of an inverse velocity. She decides to call it v_{alice} and solves for its value:

$$R_{\text{alice}}^i = v_{\text{alice}}^{-1} \frac{\ell_i}{m_i} \implies v_{\text{alice}} = \frac{\ell_i}{m_i R_{\text{alice}}^i} = 3.0 \times 10^8 \text{ m/s} \quad (3.53)$$

Alice also has colleagues that work on electromagnetism and she recognizes this value as exactly the speed of light, $v_{\text{alice}} = c$. How did that happen? She does not know, but she is surely excited about the result, as she too recognises that c is the next fundamental "speed threshold" and so is a "natural" value in the Effective Theory correction. She has explained all the planetary precession data. She writes down on a piece of paper her new theory of gravity,

$$L_{\text{alice}} = \frac{1}{2}m\dot{\mathbf{r}}^2 + \frac{GMm}{r}\left(1 + \frac{1}{c^2}\frac{\ell^2/m^2}{r^2}\right). \quad (3.54)$$

which like Bob's theory possesses symmetry and has a measure of elegance and simplicity.

As she reflects on her theory she realizes that since angular momentum is $\ell \sim mrv$, where v is the velocity of the planet orbiting the sun, the second term inside the parenthesis can be thought of as an m-independent v^2/c^2 correction to the Newtonian gravitational potential. Thus, she believes that she will be the first to show that the simple inverse-square law of Newton is corrected by factors of v^2/c^2. As the speed of the planet gets closer to the speed of light, Newton's theory begins to crack. So far the basic assumptions of spacetime symmetries—Galilean Invariance—are not breaking down, just the simple form of Newton's theory of gravity. Despite these successes of her theory, she remains slightly dissatisfied with one aspect. How can she convince herself, much less others, that her theory is better than Bob's? Surely one or the other or some combination of these corrections are required by nature, she reasons, but can they be determined from deeper theory principles? The answer is yes, and Einstein's General Relativity is that theory.

3.6.3 Gerber's "Utterly Worthless" Theory

Before going to Einstein's General Relativity, let us comment briefly on velocity dependent approaches to augmenting Newton's law. Manipulations of the Newtonian potential were initiated in earnest well after Laplace's work with the goal of rigorously incorporating finite speed effects of gravity. The most straightforward approaches failed. However, Paul Gerber proposed in 1898 (Gerber 1898) a velocity dependent potential correction that correctly accounted for Mercury's perihelion precession:

$$V(r, v) = -\frac{M}{r} \left(1 - \frac{v}{c}\right)^{-2} \tag{3.55}$$

where $c = 3 \times 10^8$ m/s is the speed of light, and v is the velocity of Mercury in the Sun-Mercury center of mass system.

Gerber's theory captured the attention of many due to its combined simplicity and effectiveness in accommodating Mercury's anomalous perihelion precession rate. For example, Mach wrote, "Only Paul Gerber [reference to 1898 paper] studying the motion of Mercury's perihelion ... did find that the speed of propagation of gravitation is the same as the speed of light" (Mach 1901). He was attacked for not giving good reasons for his theory—a topic we shall take up below—but he did provide a simple theory that worked. It was also a "natural" theory due to its utilization of c as the next fundamental speed scale of the theory.

Seventeen years after Gerber's potential, the question of Mercury's perihelion precession was resolved powerfully by Einstein's GR (Wald 1984). At low velocities the first-order correction to gravitational attraction of Gerber's theory matches the first-order correction of Einstein's theory. However, Einstein's approach had coherent principles and unassailable logic, and thoughts about Gerber's theory quickly faded away.

Despite the success in accommodating Mercury's perihelion precession, Gerber was roundly criticized for his theory. The strength of the reaction that Gerber faced seems harsh for somebody who actually did write down a simple theory of no free parameters with the speed of light in it that worked. It is as though the deep thinkers at the time knew there was something appealing about Gerber's work, but could not quite put their finger on it, and so harshly criticized it as a community building exercise to dismiss that kind of apparently principle-less approach to physics.

Einstein, commenting on Gerber's theory well after he had developed his own theory of General Relativity summarized the attitudes well: "But specialists in the field agree not only that Gerber's derivation is thoroughly incorrect, but that the formula cannot even be obtained as a consequence of Gerber's leading assumptions. Mr. Gerber's paper is therefore *utterly worthless*" (Capria 1999) (italics are mine). This appears to be an overly strong dismissal of Gerber's simple theory that gained so much attention.

Pauli, in his famous Encyclopedia article on Relativity said,

> Recently, an earlier attempt by P. Gerber has been discussed which tries to explain the perihelion advance of Mercury with the help of the finite velocity of propagation of gravitation, but which must be considered *completely unsuccessfully* from a theoretical point of view. For while it leads admittedly to the correct formula—though on the basis of false deductions—it must be stressed that, even so, only the numerical factor was new. (Paul 1981) (italics mine)

Whatever can be said of Gerber and his theory and the faulty logic behind his theory, it was not "utterly worthless" or "completely unsuccessful". I believe it was a crude attempt at effective theory analysis. It was something he may have intuited but was unsuccessful in articulating well due to the mindset and style of physics of the day. Back then, no term was allowed to augment a theory without it being derived first from a deeper principle. The standard rigor of the day was that laws were exact by argument and deduction, and any deviations or changes must be accounted for by a replacing new principle.

An excellent example of this prevailing attitude is provided by Max Born in his book on Einstein's theory of relativity (Born 1924). He describes briefly the case of Mercury's anomalous perihelion precession and then goes on to harangue all those people before Einstein who generated ad hoc solutions to the problem:

> Changes in the laws [Newton's laws] have been proposed, but they have been invented quite arbitrarily and can be tested by no other facts, and their correctness is not proved by accounting for the motion of Mercury's perihelion. If Newton's theory really requires a refinement we must demand that it emanate, without the introduction of arbitrary constants, from a principle that is superior to the existing doctrine in generality and intrinsic probability. Einstein was the first to succeed in doing this.

This attitude is partially in conflict with our understanding of Effective Theories today. The introduction of arbitrary constants is a key step in the construction of Effective Theories, and the role of experiment is to pin those down. If anything, the ad hoc inventors of changes in Newton's law were too sheepish about introducing arbitrary parameters, and instead got tangled up with incoherent "deep reasons" for their particular laws.

Effective Theory is an intermediate step between an old regime (e.g., Newton's laws) and a new regime (e.g., Einstein's General Relativity), and this intermediate step necessarily has "arbitrary couplings" and does not "emanate from a principle that is superior to the existing doctrine". Instead, it says that the existing doctrine should be taken to its utmost seriousness (e.g., Galilean invariance) and data should fit the parameters of all allowed interactions, and perhaps a deeper new theory can come along later to explain the relations among those parameters.

Although Gerber's theory was not worthless, it is not as valuable as Einstein's General Relativity. Alice and Bob's effective theories would not have been worthless either had they written it down much earlier. They would have been an intermediate stepping stone from one principled theory to the next that would have predicted the existence of Mercury's perihelion precession and motivated earlier discovery of the phenomena.

3.7 Perturbation from General Relativity

We have talked about Einstein's General Relativity being the deeper theory that explains Mercury's perihelion precession. It is worthwhile in these lectures to go through that computation to see how it comes about.

We wish to compute the trajectory of a particle subject to a central, radially symmetric gravitating source in the general approach followed, for example, by (Hartl 2003). The metric applicable for this computation is the Schwarzschild metric:

$$ds^2 = -\eta(r)c^2dt^2 + \frac{dr^2}{\eta(r)} + r^2 d\theta^2 + r^2 \sin^2\theta d\phi^2 \tag{3.56}$$

where

$$\eta(r) = 1 - \frac{2GM}{c^2 r} = 1 - \frac{r_s}{r}, \quad \text{where } r_s \equiv 2GM/c^2 \tag{3.57}$$

The quantity r_s is the Schwarzschild radius. This defines the metric tensor to be

$$g_{\alpha\beta} = \text{diag}(-\eta(r), \eta(r)^{-1}, r^2, r^2 \sin^2\theta) \tag{3.58}$$

in the (t, r, θ, ϕ) basis. Note that the signature of the metric (asymptotically weak field far away) in normal rectilinear coordinates is $g^{\alpha\beta} = \text{diag}(-1, 1, 1, 1)$.

The Schwarzschild metric is unperturbed by making shifts in the time direction and by making shifts in the angular direction ϕ. These define Killing vectors $\xi^\lambda_{\text{time}} = (1, 0, 0, 0)$ and $\xi^\lambda_{\text{rot}} = (0, 0, 0, 1)$. The nice property of a Killing vector is that when dotted into the four-velocity vector $dx^\alpha/d\tau$ the result must be constant along the geodesic motion:

$$\xi^\lambda \frac{dx_\lambda}{d\tau} = g_{\alpha\beta}\xi^\alpha \frac{dx^\beta}{d\tau} = \text{const.} \tag{3.59}$$

Applying this theorem to the Schwarzschild metric gives

$$g_{\alpha\beta}\xi^\alpha_{\text{time}} \frac{dx^\beta}{d\tau} = \eta(r)\frac{dt}{d\tau} = c_1 \tag{3.60}$$

$$g_{\alpha\beta}\xi^\alpha_{\text{rot}} \frac{dx^\beta}{d\tau} = r^2 \sin^2\theta \frac{d\phi}{d\tau} = c_2 \tag{3.61}$$

where c_1 and c_2 are mere constants. We know that independence of time implies conservation of energy, and we also know that independence of rotation implies conservation of angular momentum. Thus, we know that c_1 is some function of energy, and we know that c_2 is some function of angular momentum as we usually define the quantities. However, at this stage we do not know the precise correspondence, so it is prudent to just carry the constants c_1 and c_2 with us until the precise relations become obvious.

From Eq. 3.61 we solve for $dt/d\tau = c_1/\eta(r)$ and $d\phi/d\tau = c_2/r^2 \sin^2\theta$. Now, we should simplify this all by taking the orbit in the $\theta = \pi/2$ plane and so $d\phi/d\tau = c_2/r^2$. Please note, conservation laws have given us this, and this is where deep physics lies. Now, let's expand out the defining equation of the four-velocity

$$g_{\alpha\beta} \frac{dx^\alpha}{d\tau} \frac{dx^\beta}{d\tau} = -1, \quad \text{which gives} \tag{3.62}$$

$$-\eta(r)\left(\frac{dt}{d\tau}\right)^2 + \frac{1}{\eta(r)}\left(\frac{dr}{d\tau}\right)^2 + r^2\left(\frac{d\phi}{d\tau}\right)^2 = -1 \tag{3.63}$$

for the Schwarzschild metric. Substituting the values of $d\phi/d\tau$ and $dt/d\tau$ that we obtained above from the Killing equations, we find

$$-\frac{c_1^2}{\eta(r)} + \frac{1}{\eta(r)}\left(\frac{dr}{d\tau}\right)^2 + \frac{c_2^2}{r^2} = -1 \tag{3.64}$$

After carrying out some algebra one finds

$$\frac{mc^2}{2}(c_1^2 - 1) = \frac{1}{2}mc^2\left(\frac{dr}{d\tau}\right)^2 - \frac{GMm}{r} + \frac{mc^2c_2^2}{2r^2} - \frac{GMmc_2^2}{r^3} \tag{3.65}$$

The form of Eq. 3.65 is very suggestive of our equation for energy of a particle in an orbit, and the correspondence becomes precise if we make the identifications

$$\frac{mc^2}{2}(c_1^2 - 1) \equiv E \quad \text{and} \quad c_2^2 \equiv \frac{\ell^2}{m^2c^2}. \tag{3.66}$$

We also can identify $\tau = ct$ in the non-relativistic limit. It turns out that this substitution is acceptable for the problem at hand as long as $\dot{r} \ll r\dot{\phi}$, which is generally the situation for low eccentricity orbits, and certainly the case for the planetary orbits of our solar system. Making these identifications the energy equation becomes

$$E = \frac{1}{2}m\left(\frac{dr}{dt}\right)^2 + \frac{\ell^2}{2mr^2} - \frac{GMm}{r}\left(1 + \frac{\ell^2/m^2c^2}{r^2}\right). \qquad (3.67)$$

This is the energy equation for a particle in Newtonian gravity except for the small shift in the effective potential

$$\Delta V_{eff}(r) = -\frac{GMm}{r}\left(\frac{\ell^2/m^2c^2}{r^2}\right) \qquad (3.68)$$

which is precisely the same correction to Newton's theory we derived earlier from Alice's effective theory approach to explain Mercury's precesion in Eq. 3.54.

There are multiple ways to derive the correction to Newton's gravity law for the particular problem of perihelion precessions. In our derivation, we found Alice's theory correction. This is also the result derived in General Relativity by many other authors (see e.g., Schutz 1990; Goldstein et al. 2002; Hartl 2003). However, another approach to the General Relativity derivation gives Bob's theory, and that has been demonstrated by a set of different authors (see e.g., Paul 1981; Landau and Lifshitz 1975; Iwasaki 1971; Donoghue 2009). These two theories, if treated as god-given complete theories, are not equivalent. However, they are equivalent results for this problem as all approximations and culling of the General Relativity terms have been carried out with the sole purpose of finding the perihelion precession. In the end, the precession rate angle per orbit period from either correction is the same:

$$\delta = \frac{6\pi GM/c^2}{a(1 - e^2)} \qquad (3.69)$$

Algebraically, the orbital identity

$$\ell^2 = GMm^2a(1 - e^2). \qquad (3.70)$$

is what guarantees that the two solutions predict the same anomalous perihelion precession rate. So, we see that Albert explains both Alice's theory and Bob's theory, and puts them on firmer footing.

3.8 Conclusions

At the beginning of these lectures we decided that Newton's law of Gravitation was very successful in describing the orbits, but that it is not the precise law that captures our most profound admiration. Rather, it is the symmetries that the theory possesses. We elevated those to the highest principles and constructed reasonable effective theories that could be expected by the data. We illustrated the results with two theories: Bob's $1/r^2$ and Alice's $1/r^3$ potential correction theories. Both theories were able to account for the perihelion rate naturally. We even made the case that philosophical challenges to Newton's world view, if taken seriously, could presage the size of Mercury's correction that was actually measured by Le Verrier. This is done with the aid of "naturalness" arguments about the speed of light being the next speed scale of nature by which to construct corrections to Newton's potential. In this way the concepts of natural effective theory have some predictive power. That power is certainly qualitative, but also to some degree quantitative.

Einstein had keen insights into the nature of space and time and developed the theory of General Relativity based on them. It describes gravity at a deeper level, and one of its first orders of business was to compute the anomalous precession rate of Mercury to see if it could account for the discrepancy between Newton's theory and measurement. The answer is yes, and we have shown that this correction matches nicely the effective theory results of Alice and Bob.

Einstein's General Relativity theory is "better" than Alice's theory or Bob's theory for two reasons. First, it gives a deeper principles understanding of the correction with no additional free parameters. This deeper understanding is nothing other than further assumptions on spacetime symmetries that panned out. Second, Einstein's theory is a more complete theory of gravity that makes additional predictions (such as bending of light, and binary pulsar spin-down) that are confirmed by data. Alice or Bob's theory clearly cannot match the riches of General Relativity and so cannot be considered as fundamental as Einstein's.

Despite Bob and Alice's theory coming up short, the general lesson remains. Newton's theory was an effective theory, which is in some aspects superceded in success by Bob and Alice's effective theory, and Bob and Alice's effective theories are superceded in success by Einstein's General Relativity. The obvious next question is whether Einstein's General Relativity theory can be succeeded in success by another theory. A deeper theory that perhaps could be explained as effective theory expansion of Einstein's theory for the purposes of solving some lower energy precision measurement problem. There is little doubt that is the case (Donoghue 1994).

Finally, one of the most profound shifts in our thinking over the decades, illustrated well by the Perihelion precession example, is that it is really no longer appropriate to speak of "the correct theory." There is no correct theory. Our tasks are to improve theories via the effective theory approach, to seek deeper and simplifying assumptions that account for it, solidify those into a new theory, and then treat that new theory as an effective theory, and repeat. These steps are accomplished by continually improving and refining observations and theory computations that enable us to choose between effective theories, followed by deducing deeper new symmetries that force its inevitability. Theories are never to be trusted—they are always "wrong" in the end—and with concerted effort we can even anticipate when and how they will break down.

The concepts of Effective Theory lead one to predict qualitatively that a perihelion precession of Mercury was *a priori* guaranteed even knowing only the experimental facts of the Newtonian era. In particular, elevating symmetries above the concretization of hypothesized law, in this case the rigid devotion to the inverse square law, is the basic ingredient that would have led unambiguously to this conclusion. The general approach to science during the Newtonian era required almost complete devotion to concrete laws and their propositional justifications, which impeded its progress toward developing theory enhancements guided by symmetries and naturalness. Gerber, a school teacher who was perhaps not as indoctrinated in this more rigid fashion, found a potential that worked yet then made unjustified arguments for why it should be true. Effective Theories give the best of both words: deep but modest justifications for theories that can anticipate data and fit the data.

We have also shown that even during the time of Newton a reasonably well supported hypothesis for the perihelion precession of Mercury could have been put forth that is close to the actual experimental result of 43″ of arc per century. This is a clear illustration of how the ideas of Effective Theory can be utilized to extrapolate modestly beyond the rigidly set forth laws of fundamental physics.

References

Abrams, C.F.: Effective theories for materials and macromolecules, Institute for Mathematics and Its Applications, University of Minnesota (8–11 June 2005). http://www.ima.umn.edu/matter/W6.8-11.05/

Born, M.: Einstein's Theory of Relativity. Dover Publications, New York, 1962, 1965, original German edition 1920, and original English edition (1924)

Bradley, J.: Account of a new discovered motion of the fixed Stars. Phil. Trans. **35**, 637 (1729)

Capria, M.M.: Common sense, the history, and the theory of relativity. Acta Scientiarum **21**, 779 (1999)

Cohen, A.G.: Selected topics in effective field theories for particle physics. In: Proceedings of the Theoretical Advanced Study Institute (TASI 1993), Boulder, Colorado (1993)

Delgado-Bucalioni, R. et al.: Hybrid molecular-continuum fluid models. Phil. Trans. R. Soc. A **363**, 1975 (2005)

Donoghue, J.F.: General relativity as an effective field theory: the leading quantum corrections. Phys. Rev. D **50**, 3874 (1994). [arXiv:gr-qc/9405057]

Donoghue, J.F.: PoS **EFT09**, 001 (2009). [arXiv:0909.0021 [hep-ph]]

Duncombe, R.L.: Relativity effects for the three inner planets. Astron. J. **61**, 174 (1956)

Gerber, P.: Die räumliche und zeitliche Ausbreitung der Gravitation. Zeitschrift für Mathematik und Physik **43**, 93 (1898)

Goldstein, H., Poole, C., Safko, J.: Classical Mechanics, 3rd edn. Addison-Wesley, San Francisco (2002)

Hartl, J.: Gravity: An Introduction to Einstein's General Relativity. Addison-Wesley, San Francisco (2003)

Iwasaki, Y.: Quantum theory of gravitation vs. classical theory. Prog. Theor. Phys. **46**, 1587 (1971)

Kant, I.: *Gedanken von der wahren Schätzung der lebendigen Kräfte*, M.E. Dorn, Königsberg (1747)

Landau, L.D. Lifshitz, E.M.: The Classical Theory of Fields, 4th revised English edn. Pergamon Press, Oxford (1975)

Laplace, P.S.: Traité de Mécanique Céleste. Tome III, Paris, 1797–1825 (1805)

Le Verrier, U.: Theorie du mouvement de Mercure. Ann. Observ. Imp. Paris (Mém) **5**, 1 (1859)

Mach, E.: Die Mechanik in ihrer historisch-kritisch Entwicklung dargestellt, Leipzig (1901) (As quoted in Capria 1999)

Newcomb, S.: Entry Under "Mercury" in Encyclopedia Britannica, 11th edn. Cambridge University Press, Cambridge (1910)

Newcomb, S.: Discussion and results of observations on transits of Mercury from 1677 to 1881. Astro. Pap. Am. Ephem. Naut. Alm. **1**, 367 (1882)

Newton, I.: The Principia: Mathematical Principles of Natural Philosophy, translated by Cohen I.B., Whitman, A. 3rd edn. University of California Press, Berkeley (1999) [Newton's original three editions were published in 1687, 1713, and 1726. This book is translated from Newton's 3rd edition from 1726]

Oppenheim, S.: Kritik des newtonschen Gravitationsgesetzes. In: Encyklopädie der mathematischen Wissenschaften mit Einschluss ihrer Anwendungen, 6.2.2: 80 (1920)

Paul, W.: Theory of Relativity. Dover Publications, New York, 1958, 1981, translated from Relativitätstheorie, Encyclopädie der mathematischen Wissenschaft, vol. V19. B.G. Teubner, Leipzig (1921)

Polchinski, J.: Effective Field Theory and the Fermi Surface (1992). arXiv:hep-th/9210046

Poor, C.L.: The motions of the planets and the relativity theory. Science **54**, 30 (1921)

Rømer, O.: Touchant le mouvement de la lumiere trouvé par M. Rømer de l'Académie Royale des Sciences. Journals des Sçavans 233–236 (1676)

Roseveare, N.T.: Mercury's Perihelion from Le Verrier to Einstein. Clarendon Press, Oxford (1982)

Rothstein, I.Z.: TASI lectures on effective field theories. In: Proceedings of the Theoretical Advanced Study Institute (TASI 2003), Boulder, Colorado (2003)

Schutz, B.F.: A First Course in General Relativity. Cambridge University Press, Cambridge (1990)

Schwarzschild, K.: Über das Gravitationsfeld eines Massenpunktes nach der Einsteinschen Theorie. Klasse für Mathematik, Physik, und Technik, Berlin, p. 189 (1916)

Thayer, H.S.: Newton's Philosophy of Nature: Selections from His Writings. Haftner Press, New York (1953)

Wald, R.M.: General Relativity. University of Chicago Press, Chicago (1984)

Weinberg, S.: Effective Field Theory, Past and Future (2009). arXiv:0908.1964 [hep-th]

Will, C.M.: The Confrontation between general relativity and experiment. Living Rev. Rel. **9**, 3 (2005). [gr-qc/0510072]

Chapter 4
Effective Theories and Elementary Particle Masses

Abstract The concepts of effective theory have a rich history in particle physics. The early days of effective theories have many examples, including Fermi's theory of nucleon decay and chiral lagrangian dynamics for pion scattering. These examples are touched upon briefly before going to the most pressing issue of today, which is the origin of elementary particle masses. The problem of mass generation is first described, where it is shown that simply writing down mass terms manifestly breaks cherished symmetries. It is then shown that spontaneous symmetry breaking cures this problem. The influence of effective field theory is then addressed, where it is shown that the smallness of neutrino masses nicely conforms with our intuition, but the weak-scale value of the Higgs boson mass is confusing. The chapter concludes with an essay describing this mystery and what the resolutions might be.

4.1 Introduction

Effective theories play a central role in particle physics. Perhaps the most famous effective theory of them all is Fermi's four-fermion interaction theory that described nucleon decay and muon decay. The theory is a "V-A theory" (vector minus axial vector interaction) and has the form:

$$\mathscr{L}_{V-A} = -\frac{G_F}{\sqrt{2}} \bar{v}_\ell \gamma^\mu (1 - \gamma^5) f_\ell \, \bar{f}_q \gamma_\mu (1 - \gamma^5) f_{q'} \qquad (4.1)$$

where $G_F = 1.15 \times 10^{-6} \, \text{GeV}^{-2}$ is the Fermi constant determined by experiment. These operators can then induce β decays of the neutron via the constituent quark decays $d \rightarrow ue\bar{v}$, and can also induce muon decay through $\mu \rightarrow ev_\mu \bar{v}_e$. The history behind determining the precise nature of this interaction is a fascinating one that required painstaking experiment and insightful theory (Renton 1990).

J. D. Wells, *Effective Theories in Physics*, SpringerBriefs in Physics,
DOI: 10.1007/978-3-642-34892-1_4, © The Author(s) 2012, corrected publication 2022

We know now that the Fermi theory is just the low-energy limit of the electroweak theory of the Standard Model.[1] The Fermi constant G_F that gives the strength of the four-fermion interaction is the low-energy limit of a W-boson propagator multiplied by its couplings to the two bilinear currents:

$$\lim_{p^2 \to 0} \frac{-g^2}{p^2 - M_W^2} \implies \frac{g^2}{M_W^2} \equiv \frac{G_F}{\sqrt{2}} \tag{4.2}$$

where g is the $SU(2)_L$ gauge coupling of the Standard Model. Thus the propagator of the W-boson at very low energies compared with the W mass contracts to a point and makes an effective four-fermion interaction term governed by the Fermi Effective Theory coupling constant G_F.

Another place where Effective Theories are put to good use is in low-energy pion scattering theory. Pions are the lightest strongly interacting hadrons known in nature. The pions will interact with a very large number of other hadrons in the theory to mediate and alter even pure pion-pion scattering. Computing all of these interactions with the multitude of other intermediate hadrons is a daunting prospect to say the least. However, the effective lagrangian approach allows one to simplify these complicated dynamics of higher mass particles interactions into a few low-energy parameters of a chiral lagrangian. This technique is described well in many places (Donoghue et al. 1992).

Yet another manifestation of the power of effective theories is Wilson's discovery of the renormalization group (Wilson and Kogut 1974; Peskin and Schroeder 1995). There it was understood in a general way that at low energies all modes can be "integrated out" to form an effective lagrangian with renormalization group improved parameters. This integration-out procedure was not just hiding the effects of heavier particles into non-dynamical lagrangian mass scales, but also resuming all the higher momentum mode contributions above a cut-off scale. This technique has been extremely powerful in particle physics as both a technically useful tool to resum large quantum logarithms, but also as a conceptual tool to understand the energy flow of a theory. All modern quantum field theory textbooks, including the one listed in Wilson and Kogut (1974); Peskin and Schroeder (1995), have very thorough treatments of this most important issue.

There are numerous other examples of effective theories being employed in the particle physics context. All of the theories of physics beyond the Standard Model also utilize the concepts in one form or another. The language of effective theory concepts is so deeply ingrained in the minds of practitioners now there is rarely need to explicit point out or argue for its utility.

There is, however, one area of particle physics where the notions of effective theories are hard to mesh with reality. This is regarding the structure of the vacuum. For one, effective theory concepts would tell us that the cosmological constant is many orders of magnitude beyond what we observe today. This is usually just ignored in the field, with hope that some other quantum-gravity solution as yet not understood

---

[1] The electroweak theory of the Standard Model will be discussed in more detail in Sect. 4.3.

will come to save the cosmological constant. I will not talk about this. The second place where our experimental understanding of the vacuum may be at odds with effective theories is in the generation of elementary particle masses. That will be the focus of this chapter. I will first outline the challenges to giving mass to elementary particles in chiral theories and then I will give a brief introduction to the Standard Model electroweak theory. After that I will describe the effective theory issues for the masses of leptons and quarks, neutrinos, and the Higgs boson. The Higgs boson is especially interesting since it has likely been discovered lately (Aad et al. 2012; Chatrchyan et al. 2012), and the theoretical controversies surrounding why its mass is light are very hot today.

4.2 The Problem of Mass in Chiral Gauge Theories

The fermions of the Standard Model and some of the gauge bosons have mass. This is a troublesome statement since gauge invariance appears to allow neither. Let us review the situation for gauge bosons and chiral fermions and introduce the Higgs mechanism that solves it. First, we illustrate the concepts with a massive $U(1)$ theory—spontaneously broken QED.

Gauge Boson Mass
  The lagrangian of QED is

$$\mathscr{L}_{QED} = -\frac{1}{4}F_{\mu\nu}F^{\mu\nu} + \bar{\psi}(i\gamma^{\mu}D_{\mu} - m)\psi \qquad (4.3)$$

where

$$D_{\mu} = \partial_{\mu} + ieA_{\mu} \qquad (4.4)$$

and $Q = -1$ is the charge of the electron. This lagrangian respects the $U(1)$ gauge symmetry

$$\psi \rightarrow e^{-i\alpha(x)}\psi \qquad (4.5)$$

$$A_{\mu} \rightarrow A_{\mu} + \frac{1}{e}\partial_{\mu}\alpha(x). \qquad (4.6)$$

Since QED is a vector-like theory—left-handed electrons have the same charge as right-handed electrons—an explicit mass term for the electron does not violate gauge invariance.
  If we wish to give the photon a mass we may add to the lagrangian the mass term

$$\mathscr{L}_{mass} = \frac{m_A^2}{2}A_{\mu}A^{\mu}. \qquad (4.7)$$

However, this term is not gauge invariant since under a transformation $A_\mu A^\mu$ becomes

$$A_\mu A^\mu \rightarrow A_\mu A^\mu + \frac{2}{e} A^\mu \partial_\mu \alpha + \frac{1}{e^2} \partial_\mu \alpha \partial^\mu \alpha \qquad (4.8)$$

This is not the right way to proceed if we wish to continue respecting the gauge symmetry. There is a satisfactory way to give mass to the photon while retaining the gauge symmetry. This is the Higgs mechanism, and the simplest way to implement it is via an elementary complex scalar particle that is charged under the symmetry and has a vacuum expectation value (vev) that is constant throughout all space and time. This is the Higgs boson field Φ.

Let us suppose that the photon in QED has a mass. To see how the Higgs boson implements the Higgs mechanism in a gauge invariant manner, we introduce the field Φ with charge q to the lagrangian:

$$\mathscr{L} = \mathscr{L}_{QED} + (D_\mu \Phi)^* (D^\mu \Phi) - V(\Phi) \qquad (4.9)$$

where

$$V(\Phi) = \mu^2 |\Phi|^2 + \lambda |\Phi|^4 \qquad (4.10)$$

where it is assumed that $\lambda > 0$ and $\mu^2 < 0$.

Since Φ is a complex field we have the freedom to parametrize it as

$$\Phi = \frac{1}{\sqrt{2}} \phi(x) e^{i\xi(x)}, \qquad (4.11)$$

where $\phi(x)$ and $\xi(x)$ are real scalar fields. The scalar potential with this choice simplifies to

$$V(\Phi) \rightarrow V(\phi) = \frac{\mu^2}{2} \phi^2 + \frac{\lambda}{4} \phi^4. \qquad (4.12)$$

Minimizing the scalar potential one finds

$$\left. \frac{dV}{d\phi} \right|_{\phi=\phi_0} = \mu^2 \phi_0 + \lambda \phi_0^3 = 0 \implies \phi_0 = \sqrt{\frac{-\mu^2}{\lambda}}. \qquad (4.13)$$

This vacuum expectation value of ϕ enables us to normalize the ξ field by ξ/ϕ_0 such that its kinetic term is canonical at leading order of small fluctuation, legitimizing the parametrization of Eq. (4.11). We can now choose the unitary gauge transformation, $\alpha(x) = -\xi(x)/\phi_0$, to make Φ real-valued everywhere. One finds that the complex scalar kinetic terms expand to

$$(D_\mu \Phi)^* (D^\mu \Phi) \rightarrow \frac{1}{2} (\partial_\mu \phi)^2 + \frac{1}{2} e^2 q^2 \phi^2 A_\mu A^\mu \qquad (4.14)$$

At the minimum of the potential $\langle \phi \rangle = \phi_0$, so one can expand the field ϕ about its vev, $\phi = \phi_0 + h$, and identify the fluctuating degree of freedom h with a propagating real scalar boson.

The Higgs boson mass and self-interactions are obtained by expanding the lagrangian about ϕ_0. The result is

$$- \mathscr{L}_{Higgs} = \frac{m_h^2}{2} h^2 + \frac{\mu'}{3!} h^3 + \frac{\eta}{4!} h^4 \qquad (4.15)$$

where

$$m_h^2 = 2\lambda\phi_0^2, \quad \mu' = \frac{3m_h^2}{\phi_0}, \quad \eta = 6\lambda = 3\frac{m_h^2}{\phi_0^2}. \qquad (4.16)$$

The mass of the Higgs boson is not dictated by gauge couplings here, but rather by its self-interaction coupling λ and the vev.

The complex Higgs boson kinetic terms can be expanded to yield

$$\Delta \mathscr{L} = \frac{1}{2} e^2 q^2 \phi_0^2 A_\mu A^\mu + e^2 q^2 h A_\mu A^\mu + \frac{1}{2} e^2 q^2 h^2 A_\mu A^\mu. \qquad (4.17)$$

The first term is the mass of the photon, $m_A^2 = e^2 q^2 \phi_0^2$. A massive vector boson has a longitudinal degree of freedom, in addition to its two transverse degrees of freedom, which accounts for the degree of freedom lost by virtue of gauging away $\xi(x)$. The second and third terms of Eq. 4.17 set the strength of interaction of a single Higgs boson and two Higgs bosons to a pair of photons:

$$hA_\mu A_\nu \text{ Feynman rule}: \ i2e^2 q^2 \phi_0 g_{\mu\nu} = i2 \frac{m_A^2}{\phi_0} \qquad (4.18)$$

$$hhA_\mu A_\nu \text{ Feynman rule}: \ i2e^2 q^2 g_{\mu\nu} = i2 \frac{m_A^2}{\phi_0^2} \qquad (4.19)$$

after appropriate symmetry factors are included.

The general principles to retain from this discussion are first that massive gauge bosons can be accomplished in a gauge-invariant way through the Higgs mechanism. The Higgs boson that gets a vev breaks whatever symmetries it is charged under—the Higgs vev carries charge into the vacuum. And finally, the Higgs boson that gives mass to the gauge boson couples to it proportional to the gauge boson mass.

Chiral Fermion Masses

In quantum field theory a four-component fermion can be written in its chiral basis as

$$\psi = \begin{pmatrix} \psi_L \\ \psi_R \end{pmatrix} \qquad (4.20)$$

where $\psi_{L,R}$ are two-component chiral projection fermions. A mass term in quantum field theory is equivalent to an interaction between the ψ_L and ψ_R components

$$m\bar{\psi}\psi = m\psi_L^{\dagger}\psi_R + m\psi_R^{\dagger}\psi_L. \tag{4.21}$$

In vectorlike QED, the ψ_L and ψ_R components have the same charge and a mass term can simply be written down. However, let us now suppose that in our toy $U(1)$ model, there exists a set of chiral fermions where the $P_L\psi = \psi_L$ chiral projection carries a different gauge charge than the $P_R\psi = \psi_R$ chiral projection. In that case, we cannot write down a simple mass term without explicitly breaking the gauge symmetry.

The resolution to this conundrum of masses for chiral fermions resides in the Higgs sector. If the Higgs boson has just the right charge, it can be utilized to give mass to the chiral fermions. For example, if the charges[2] are $Q[\psi_L] = 1$, $Q[\psi_R] = 1 - q$ and $Q[\Phi] = q$ we can form the gauge invariant combination

$$\mathcal{L}_f = y_\psi\, \psi_L^{\dagger}\Phi\psi_R + c.c. \tag{4.22}$$

where y_f is a dimensionless Yukawa coupling. Now expand the Higgs boson about its vev, $\Psi = (\phi_0 + h)/\sqrt{2}$, and we find

$$\mathcal{L}_f = m_\psi\, \psi_L^{\dagger}\psi_R + \left(\frac{m_\psi}{\phi_0}\right) h\psi_L^{\dagger}\psi_R + c.c. \tag{4.23}$$

where $m_\psi = y_\psi\phi_0/\sqrt{2}$.

We have successfully generated a mass by virtue of the Yukawa interaction with the Higgs boson. That same Yukawa interaction gives rise to an interaction between the physical Higgs boson and the fermions:

$$h\bar{\psi}\psi \;(\text{Feynman rule}) \;:\; i\frac{m_\psi}{\phi_0}. \tag{4.24}$$

Just as was the case with the gauge bosons, the generation of fermion masses by the Higgs boson leads to an interaction of the physical Higgs bosons with the fermion proportional to the fermion mass. As we will see in the Standard Model, this rigid connection between mass and interaction is what enables us to anticipate Higgs boson phenomenology with great precision once the mass is precisely known.

4.3 Standard Model Electroweak Theory

The bosonic electroweak lagrangian is an $SU(2)_L \times U(1)_Y$ gauge invariant theory

$$\mathcal{L}_{bos} = |D_\mu\Phi|^2 - \mu^2|\Phi|^2 - \lambda|\Phi|^4 - \frac{1}{4}B_{\mu\nu}B^{\mu\nu} - \frac{1}{4}W_{\mu\nu}^a W^{a,\mu\nu} \tag{4.25}$$

---

[2] We ignore the additional fields that would be needed in order to make the spectrum gauge anomaly free. Doing so is straightforward and would not change the message of this example.

where Φ is an electroweak doublet with Standard Model charges of $(\mathbf{2}, 1/2)$ under $SU(2)_L \times U(1)_Y$ $(Y = +1/2)$. In our normalization electric charge is $Q = T^3 + \frac{Y}{2}$, and the doublet field Φ can be written as two complex scalar component fields ϕ^+ and ϕ^0:

$$\Phi = \begin{pmatrix} \phi^+ \\ \phi^0 \end{pmatrix}. \tag{4.26}$$

The covariant derivative and field strength tensors are

$$D_\mu \Phi = \left(\partial_\mu + ig\frac{\tau^a}{2} W_\mu^a + ig'\frac{Y}{2} B_\mu \right) \Phi \tag{4.27}$$

$$B_{\mu\nu} = \partial_\mu B_\nu - \partial_\nu B_\mu \tag{4.28}$$

$$W_{\mu\nu}^a = \partial_\mu W_\nu^a - \partial_\nu W_\mu^a - g f^{abc} W_\mu^b W_\nu^c \tag{4.29}$$

The minimum of the potential does not occur at $\Phi = 0$ if $\mu^2 < 0$. Instead, one finds that the minimum occurs at a non-zero value of Φ—its vacuum expectation value (vev)—which via a gauge transformation can always be written as

$$\langle \Phi \rangle = \frac{1}{\sqrt{2}} \begin{pmatrix} 0 \\ v \end{pmatrix} \quad \text{where} \quad v \equiv \sqrt{\frac{-\mu^2}{\lambda}}. \tag{4.30}$$

This vev carries hypercharge and weak gauge charge into the vacuum, and what is left unbroken is electric charge. This result we anticipated in Eq. (4.26) by defining a charge Q in terms of hypercharge and an eigenvalue of the $SU(2)$ generator T^3, and then writing the field Φ in terms of ϕ^0 and ϕ^+ of zero and positive $+1$ definite electric charge.

Our symmetry breaking pattern is then simply $SU(2)_L \times U(1)_Y \rightarrow U(1)_Q$. The original group, $SU(2)_L \times U(1)_Y$, has a total of four generators and $U(1)_Q$ has one generator. Thus, three generators are 'broken'. Goldstone's theorem (Goldstone et al. 1962) tells us that for every broken generator of a symmetry there must correspond a massless field. These three massless Goldstone bosons we can call $\phi_{1,2,3}$. We now can rewrite the full Higgs field Φ as

$$\langle \Phi \rangle = \frac{1}{\sqrt{2}} \begin{pmatrix} 0 \\ v \end{pmatrix} + \frac{1}{\sqrt{2}} \begin{pmatrix} \phi_1 + i\phi_2 \\ h + i\phi_3 \end{pmatrix} \tag{4.31}$$

The fourth degree of freedom of Φ is the Standard Model Higgs boson h. It is a propagating degree of freedom. The other three states $\phi_{1,2,3}$ can all be absorbed as longitudinal components of three massive vector gauge bosons Z, $W^\pm$ which are defined by

$$W_\mu^\pm = \frac{1}{\sqrt{2}} \left(W_\mu^{(1)} \mp i W_\mu^{(2)} \right) \tag{4.32}$$

$$B_\mu = \frac{-g' Z_\mu + g A_\mu}{\sqrt{g^2 + g'^2}} \tag{4.33}$$

$$W_\mu^{(3)} = \frac{g Z_\mu + g' A_\mu}{\sqrt{g^2 + g'^2}}. \tag{4.34}$$

It is convenient to define $\tan \theta_W = g'/g$. By measuring interactions of the gauge bosons with fermions it has been determined experimentally that $g = 0.65$ and $g' = 0.35$, and therefore $\sin^2 \theta_W = 0.23$.

After performing the redefinitions of the fields above, the kinetic terms for the $W_\mu^\pm$, Z_μ, A_μ will all be canonical. Expanding the Higgs field about the vacuum, the contributions to the lagrangian involving Higgs boson interaction terms are

$$\mathcal{L}_{h\,int} = \left[m_W^2 W_\mu^+ W^{-,\mu} + \frac{m_Z^2}{2} Z_\mu Z^\mu \right] \left(1 + \frac{h}{v} \right)^2 \tag{4.35}$$

$$- \frac{m_h^2}{2} h^2 - \frac{\xi}{3!} h^3 - \frac{\eta}{4!} h^4 \tag{4.36}$$

where

$$m_W^2 = \frac{1}{4} g^2 v^2, \quad m_Z^2 = \frac{1}{4}(g^2 + g'^2) v^2 \implies \frac{m_W^2}{m_Z^2} = 1 - \sin^2 \theta_W \tag{4.37}$$

$$m_h^2 = 2\lambda v^2, \quad \xi = \frac{3 m_h^2}{v}, \quad \eta = 6\lambda = \frac{3 m_h^2}{v^2}. \tag{4.38}$$

From our knowledge of the gauge couplings, the value of the vev v can be determined from the masses of the gauge bosons: $v \simeq 246\,\mathrm{GeV}$.

The Feynman rules for Higgs boson interactions are

$$hhh : -\frac{i3 m_h^2}{v} \tag{4.39}$$

$$hhhh : -i \frac{3 m_h^2}{v^2} \tag{4.40}$$

$$h W_\mu^+ W_\nu^- : i2 \frac{m_W^2}{v} g^{\mu\nu} \tag{4.41}$$

$$h Z_\mu Z_\nu : i2 \frac{m_Z^2}{v} g_{\mu\nu} \tag{4.42}$$

$$hh W_\mu^+ W_\nu^- : i2 \frac{m_W^2}{v^2} g_{\mu\nu} \tag{4.43}$$

$$hh Z_\mu Z_\nu : i2 \frac{m_Z^2}{v^2} g_{\mu\nu} \tag{4.44}$$

Fermion masses are also generated in the Standard Model through the Higgs boson vev, which in turn induces an interaction between the physical Higgs boson and the fermions. Let us start by looking at b quark interactions. The relevant lagrangian for couplings with the Higgs boson is

$$\Delta \mathscr{L} = y_b Q_L^\dagger \Phi b_R + c.c. \quad \text{where} \quad Q_L^\dagger = (t_L^\dagger \ b_L^\dagger) \tag{4.45}$$

where y_b is the Yukawa coupling. The Higgs boson, after a suitable gauge transformation, can be written simply as

$$\Phi = \frac{1}{\sqrt{2}} \begin{pmatrix} 0 \\ v+h \end{pmatrix} \tag{4.46}$$

and the interaction lagrangian can be expanded to

$$\Delta \mathscr{L} = y_b Q_L^\dagger \Phi b_R + c.c. = \frac{y_b}{\sqrt{2}} (t_L^\dagger \ b_L^\dagger) \begin{pmatrix} 0 \\ v+h \end{pmatrix} b_R + h.c. \tag{4.47}$$

$$= m_b (b_R^\dagger b_L + b_L^\dagger b_R) \left(1 + \frac{h}{v}\right) = m_b \bar{b} b \left(1 + \frac{h}{v}\right) \tag{4.48}$$

where $m_b = y_b v / \sqrt{2}$ is the mass of the b quark.

The quantum numbers work out perfectly to allow this mass term. See Table 4.1 for the quantum numbers of the various fields under the Standard Model symmetries. Under $SU(2)$ the interaction $Q_L^\dagger \Phi b_R$ is invariant because $\mathbf{2} \times \mathbf{2} \times \mathbf{1} \in \mathbf{1}$ contains a singlet. And under $U(1)_Y$ hypercharge the interaction is invariant because $Y_{Q_L^\dagger} + Y_\Phi + Y_{b_R} = -\frac{1}{6} + \frac{1}{2} - \frac{1}{3}$ sums to zero. Thus, the interaction is invariant under all gauge groups, and we have found a suitable way to give mass to the bottom quark.

How does this work for giving mass to the top quark? Obviously, $Q_L^\dagger \Phi t_R$ is not invariant. However, we have the freedom to create the conjugate representation of Φ which still transforms as a $\mathbf{2}$ under $SU(2)$ but switches sign under hypercharge: $\Phi^c = i\sigma^2 \Phi^*$. This implies that $Y_{\Phi^c} = -\frac{1}{2}$ and

$$\Phi^c = \frac{1}{\sqrt{2}} \begin{pmatrix} v+h \\ 0 \end{pmatrix} \tag{4.49}$$

when restricted to just the real physical Higgs field expansion about the vev. Therefore, it becomes clear that $y_t Q_L^\dagger \Phi^c t_R + c.c.$ is now invariant since the $SU(2)$ invariance remains $\mathbf{2} \times \mathbf{2} \times \mathbf{1} \in \mathbf{1}$ and $U(1)_Y$ invariance follows from $Y_{Q_L^\dagger} + Y_{\Phi^c} + Y_{t_R} = -\frac{1}{6} - \frac{1}{2} + \frac{2}{3} = 0$. Similar to the b quark one obtains an expression for the mass and Higgs boson interaction:

Table 4.1 Charges of standard model fields

Field	$SU(3)$	$SU(2)_L$	T^3	$\dfrac{Y}{2}$	$Q = T^3 + \dfrac{Y}{2}$
g_μ^a (gluons)	8	1	0	0	0
$(W_\mu^\pm, W_\mu^0)$	1	3	$(\pm 1, 0)$	0	$(\pm 1, 0)$
B_μ^0	1	1	0	0	0
$Q_L = \begin{pmatrix} u_L \\ d_L \end{pmatrix}$	3	2	$\begin{pmatrix} \frac{1}{2} \\ -\frac{1}{2} \end{pmatrix}$	$\frac{1}{6}$	$\begin{pmatrix} \frac{2}{3} \\ -\frac{1}{3} \end{pmatrix}$
u_R	3	1	0	$\frac{2}{3}$	$\frac{2}{3}$
d_R	3	1	0	$-\frac{1}{3}$	$-\frac{1}{3}$
$E_L = \begin{pmatrix} \nu_L \\ e_L \end{pmatrix}$	1	2	$\begin{pmatrix} \frac{1}{2} \\ -\frac{1}{2} \end{pmatrix}$	$-\frac{1}{2}$	$\begin{pmatrix} 0 \\ -1 \end{pmatrix}$
e_R	1	1	0	-1	-1
$\Phi = \begin{pmatrix} \phi^+ \\ \phi^0 \end{pmatrix}$	1	2	$\begin{pmatrix} \frac{1}{2} \\ -\frac{1}{2} \end{pmatrix}$	$\frac{1}{2}$	$\begin{pmatrix} 1 \\ 0 \end{pmatrix}$
$\Phi^c = \begin{pmatrix} \phi^0 \\ \phi^- \end{pmatrix}$	1	2	$\begin{pmatrix} \frac{1}{2} \\ -\frac{1}{2} \end{pmatrix}$	$-\frac{1}{2}$	$\begin{pmatrix} 0 \\ -1 \end{pmatrix}$

$$\Delta \mathcal{L} = y_t Q_L^\dagger \Phi^c t_R + c.c. = \frac{y_t}{\sqrt{2}} (t_L^\dagger \ b_L^\dagger) \begin{pmatrix} v + h \\ 0 \end{pmatrix} t_R + c.c. \tag{4.50}$$

$$= m_t (t_R^\dagger t_L + t_L^\dagger t_R) \left(1 + \frac{h}{v} \right) = m_t \, \bar{t} t \left(1 + \frac{h}{v} \right) \tag{4.51}$$

where $m_t = y_t v / \sqrt{2}$ is the mass of the t quark.

The mass of the charged leptons follows in the same manner, $y_e E_L^\dagger \Phi e_R + c.c.$, and interactions with the Higgs boson result. In all cased the Feynman diagram for Higgs boson interactions with the fermions at leading order is

$$h \bar{f} f \; : \; i \frac{m_f}{v}. \tag{4.52}$$

We see from this discussion several important points. First, the single Higgs boson of the Standard Model can give mass to all Standard Model states, even to the neutrinos as we will see in the next section. It did not have to be that way. It could have been that quantum numbers of the fermions did not enable just one Higgs boson to give mass to everything. This is the Higgs boson miracle of the Standard Model. The second thing to keep in mind is that there is a direct connection between the Higgs boson giving mass to a particle and it interacting with that particle. We have seen that all interactions are directly proportional to a mass factor. This is why Higgs boson phenomenology is completely determined in the Standard Model just from the Higgs boson mass.

4.4 The Special Case of Neutrino Masses

For many years it was thought that neutrinos might be exactly massless. Although recent experiments have shown that this is not the case, the masses of neutrinos are extraordinarily light compared to other Standard Model fermions. In this section we discuss the basics of neutrino masses (Grossman 2003; De Gouvea 2004; Mohapatra 2004; Altarelli 2007), with emphasis on how the Higgs boson plays a role.

Some physicists define the Standard Model without a right-handed neutrino. Thus, there is no opportunity to write down a Yukawa interaction of the left and right-handed neutrinos with the Higgs boson that gives neutrinos a mass. A higher-dimensional operator is needed,

$$\mathcal{O}_v = \frac{\lambda_{ij}}{\Lambda}(E_{iL}^\dagger H^c)^\dagger (E_{jL}^\dagger H^c) \tag{4.53}$$

where $E_L = (v_L \; e_L)$ is the $SU(2)$ doublet of left-handed neutrino and electron. Taking into account the various flavors $i = 1, 2, 3$ results in a 3×3 mass matrix for neutrino masses

$$(m_v)_{ij} = \lambda_{ij} \frac{v^2}{\Lambda}. \tag{4.54}$$

Λ can be considered the cutoff of the Standard Model effective theory (see Sect. 4.5), and the operator given by Eq. (4.53) is the only gauge-invariant, Lorentz-invariant operator that one can write down at the next higher dimension ($d = 5$) in the theory. Thus, it is a satisfactory approach to neutrino physics, leading to an indication of new physics beyond the Standard Model at the scale Λ. For this reason, many view the existence of neutrino masses as a signal for physics beyond the Standard Model.

The absolute value of neutrino masses has not been measured but the differences of mass squareds between various neutrino masses have been measured and range from about 10^{-5} to 10^{-2} eV^2 (Grossman 2003; De Gouvea 2004; Mohapatra 2004; Altarelli 2007; Kayser 2012). It is reasonable therefore to suppose that the largest neutrino mass in the theory should be around 0.1 eV. If we assume that this mass scale is obtained using the natural value of $\lambda \sim 1$ in Eq. (4.54) and a large mass scale Λ, this sets the scale of the cutoff Λ to be

$$\Lambda \simeq \frac{(246\,\text{GeV})^2}{0.1\,\text{eV}} \simeq 10^{15}\,\text{GeV} \tag{4.55}$$

This is a very interesting scale, since it is within an order of magnitude of where the three gauge couplings of the Standard Model come closest to meeting, which may be an indication of grand unification. The scale Λ could then be connected to this Grand Unification scale.

Another approach to neutrino masses is to assume that there exists a right-handed neutrino ν_R. After all, there is no strong reason to banish this state, especially since there is an adequate right-handed partner state to all the other fermions. Furthermore, if the above considerations are pointing to a grand unified theory, right-handed neutrinos are generally present in acceptable versions, such as $SO(10)$ where all the fermions are in the **16** representation, including ν_R. Quantum number considerations indicate that ν_R is a pure singlet under the Standard Model gauge symmetries, and thus we have a complication in the neutrino mass sector beyond what we encountered for the other fermions of the theory. In particular, we are now able to add a Majorana mass term $\nu_R^T i \sigma^2 \nu_R$ that is invariant all by itself without the need of a Higgs boson. The full mass interactions available to the neutrino are now

$$\mathcal{L}_\nu = y_{ij} E_{iL}^\dagger \Phi^c \nu_{jR} + \frac{M_{ij}}{2} \nu_{iR}^T i \sigma^2 \nu_{jR} + c.c. \tag{4.56}$$

The resulting 6×6 mass matrix in the $\{\nu_L, \nu_R^c\}$ basis is

$$m_\nu = \begin{pmatrix} 0 & m_D \\ m_D^T & M \end{pmatrix} \tag{4.57}$$

where M is the matrix of Majorana masses with values M_{ij} taken straight from Eq. (4.56), and m_D are the neutrino Dirac mass matrices taken from the Yukawa interaction with the Higgs boson

$$(m_D)_{ij} = \frac{y_{ij}}{\sqrt{2}} v. \tag{4.58}$$

Consistent with effective field theory ideas, there is no reason why the Majorana mass matrix entries should be tied to the weak scale. They should be of order the cutoff scale of when the Standard Model is no longer considered complete. Therefore, it is reasonable and expected to assume that M_{ij} entries are generically much greater than the weak scale. In that limit, the seesaw matrix of Eq. (4.57) has three heavy eigenvalues of $\mathcal{O}(M)$, and three light eigenvalues that, to leading order and good approximation, are eigenvalues of the 3×3 matrix

$$m_\nu^{\text{light}} = -m_D^T M^{-1} m_D \sim y^2 \frac{v^2}{M} \tag{4.59}$$

which is parametrically of the same form as Eq. (4.54). This is expected since the light eigenvalues can be evaluated from the operators left over after integrating out the heavy right-handed neutrinos in the effective theory. That operator is simply Eq. (4.53), where schematically Λ can be associated with the scale M, and λ can be associated with y^2.

We will emphasize in the next section that the story of neutrino masses conforms very nicely with our notions of effective field theories. It is for this reason that most physicists are not terribly alarmed about the smallness of neutrino masses, even though on the surface it would appear quite disturbing to know that neutrinos are orders of magnitude in mass below other particles that we measure very directly at colliders. They are 12 orders of magnitude below the top quark mass, for example. Nevertheless, there is no concern.

The role of effective theory becomes much more troublesome to understand in the context of Higgs boson physics, even though the Higgs boson mass is in the close neighborhood (i.e., less than an order of magnitude difference) of the W, Z, and top quark masses. The effective theory issues surrounding the peculiar spin zero Higgs boson, the main focus of this chapter that we have been building to, is something we come to now.

4.5 Natural Effective Theories, the Higgs Boson, and the Hierarchy Problem

The Standard Model with its postulated Higgs boson is an unsatisfactory theory for many reasons. There are several direct data-driven reasons why it is incomplete. The Standard Model has no explanation for the baryon asymmetry of the Universe. For some reason there are many more protons than anti-protons, and if the Universe is cooling from some primordial hot state with particles in thermal equilibrium, that is unexpected. Some mechanism that goes beyond the Standard Model dynamics must be at play. Similarly, there is plenty of astrophysical evidence for dark matter in the Universe. This dark matter helps to explain structure formation, details of the cosmic microwave background radiation, galactic rotation curves, etc. The problem is the Standard Model has no candidate explanation, and new physics must be invoked.

There are many other reasons to consider physics beyond the Standard Model. The three gauge forces could be unified and the matter unified within representations of a grand unified symmetry. The many different parameters of the flavor sector are hard to swallow without envisaging deeper principles that organize them. Furthermore, the integration of the Standard Model with quantum gravity is not obvious, and many think a deeper structure, such as that built from strings and branes, is needed for their coexistence.

So, there are many reasons to believe that there is physics beyond the Standard Model. But the issue that is front and center for us now, relevant to Higgs boson physics and electroweak explorations at the Large Hadron Collider, is the Hierarchy Problem. The Hierarchy Problem is often expressed as a question: Why is the weak scale ($\sim 10^2$ GeV) so much lighter than the Planck scale ($\sim 10^{18}$ GeV)? It is a bit uninspiring when phrased this way, since it begs the question of why we should be concerned at all about a big difference in scales. Blue whales are much bigger than nanoarchaeum equitans but we do not believe nature must reveal a dramatic new concept for us to understand it (Clauset 2012).

A knowing-just-enough-to-be-dangerous naive way to look at the Standard Model is that it is the "Theory of Particles", valid up to some out-of-reach scale where gravity might go strong, or some other violence is occurring that we do not care about. It is a renormalizable theory. I can compute everything at multiple quantum loop order, set counter terms, cancel infinities that are fake since they do not show up in observables, and then make predictions for observables that experiment agrees with. Quadratic divergences of the Higgs boson self-energy, which so many people make a fuss about, are not even there if I use dimensional regularization. The theory is happy, healthy, stable, and in no need of any fixes. New physics *near the electroweak scale* can still be justified (Wells 2003, 2005; Arkani-Hamed and Dimopoulos 2005; Giudice and Romanino 2004; Arkani-Hamed et al. 2005) after dismissing naturalness as impossibly imprecise to understand at this stage, but the urgency is certainly diminished for it being *at the electroweak scale*.

This viewpoint that the Standard Model is complete can be challenged right at the outset. It is simply not the "Theory of Particles"—it does break down. It is an effective theory, even if one thinks there is a way to argue it being valid to some very remote high scale where gravity goes strong, such as M_{Pl}. As an effective theory, all operators should have their dimensionality set by the cutoff of the theory (Polchinski 1992). If operator $\mathcal{O}^{(d)}$ has dimension d then its coefficient is $c\Lambda^{4-d}$, where Λ is the cutoff of the theory and c is expected to be ~ 1 in value. Irrelevant operators with $d > 4$ cause no harm. Same goes for $d = 4$ marginal operators. The Standard Model is almost exclusively a theory of $d = 4$ marginal operators with its kinetic terms, gauge interaction terms, and Yukawa interaction terms. What is potentially problematic is the existence of any $d < 4$ relevant operators. In that case, the coefficients should be large, set by the cutoff of the theory.

Does the Standard Model have any gauge-invariant, Lorentz-invariant relevant $d < 4$ operators to worry about? Yes, two of them. The right-handed neutrino Majorana mass interaction terms $v_R^T i \sigma^2 v_R$, which is $d = 3$, and the Higgs boson mass operator $|H|^2$, which is $d = 2$. The expectations of effective field theories is that the scale of the coefficients of these operators should be set by high-scale cutoffs of the theory and disconnected from any other surviving mass scale in the infrared. As we saw in Sect. 4.4 this expectation is nicely met in the neutrino case, where we have actually measured the masses and see a self-consistent picture for large Majorana masses for the right-handed neutrinos, which serve as cutoff scale coefficients. These coefficients are tied to lepton number violation, for example, and not electroweak symmetry breaking, and therefore have naturally large values above the weak scale.

It did not have to be that way with neutrino physics. It could have been that the neutrino sector was shown experimentally to have independent left and right-handed components and the masses were of order the weak scale. This would have been in violation of effective field theory expectations, unless new symmetries tied to the weak scale were discovered to protect the right-handed neutrino from getting a large Majorana mass. The fact that the neutrino sector conforms with effective field theory expectations should be viewed as contributing evidence for these concepts.

In contrast to the neutrino operator, the $d = 2$ Higgs mass operator in the Standard Model is unwelcome if its coefficient is not set to the weak scale. From our effective field theory expectations, the Lagrangian operator should be

$$\Delta \mathscr{L}_{rel} = c \Lambda^2 |H|^2. \qquad (4.60)$$

This is a potential disaster for the theory, since from our previous work on the Higgs potential we stated that the Higgs mass must be $-\mu^2 \sim v^2$, where $v \simeq 246\,\text{GeV}$ is the Higgs boson vacuum expectation value needed to reproduce the W and Z masses. If we assume the Standard Model to be a valid theory to very high energies $E \gg v$, that implies the cutoff of the Standard Model effective theory is $\Lambda \gg v$, which "incorrectly" implies the coefficient of $|H|^2$ is $|\mu^2| = \Lambda^2 \gg v^2$. The effective theory would then need the coefficient c in Eq. (4.60) to be finetuned to an extraordinarily small and unnatural (Giudice 2004) value $c \sim v^2/\Lambda^2$ to make all the scales work out properly. The concern about how this can be so is the Hierarchy Problem.

The discussion is a bit abstract, but it bears fruit with direct computations. As one example out of an infinite number that would demonstrate the Hierarchy Problem, consider the possible existence of other scalar fields ϕ_i at higher energies. The assumption is that if there is a Higgs boson in the theory, then there is every reason to believe that there can be other scalars. They can have mass at the weak scale, intermediate scale, Planck scale, wherever. Let us suppose that we put one ϕ at the cutoff scale Λ of the theory. The operator $|\phi|^2|H|^2$ immediately gives a quantum correction to the Higgs mass operator coefficient of $\sim \Lambda^2/16\pi^2$. Although the $1/16\pi^2$ can help a little, if $\Lambda \gg 4\pi v$ there is serious problem, and the weak scale cannot exist naturally with such a hierarchy. For this reason, it is often assumed that naturalness of the Higgs boson sector of the Standard Model effective theory requires new physics to show up at some scale below $\Lambda \sim 4\pi v \sim$ few TeV.

There are many different approaches to solving the Hierarchy Problem. One approach suggests that there is new physics at the TeV scale and the cutoff Λ in Eq. (4.60) is in the neighborhood of the weak scale. Supersymmetry (Martin 1997), little Higgs (Schmaltz and Smith 2005), conformal theories (Frampton and Vafa 1999), and extra dimensions (Sundrum 2005; Rattazzi 2006) can be employed in this approach. For example, supersymmetry accomplishes the task by a softly broken symmetry, where Λ is the supersymmetry breaking mass scale. All quadratic divergences to the Higgs boson mass operator cancel up to supersymmetry breaking terms. Extra dimensions accomplishes it by banishing all mass scales accessible to the Higgs boson above the TeV scale. Another approach suggests that fundamental scalars are banished from the theory that could form invariant $|\varphi|^2$ operators. For example, this is the approach of Technicolor (Lane and Martin 2009) and top-quark condensate theories (Hill 1991, 1995; Martin 1997; Chivukula et al. 1999) that try to reproduce the symmetry breaking of a Higgs boson with the condensate of a fermion bilinear operator. Higgsless theories and their variants are also in this category (Csaki et al. 2004a,b; Cui et al. 2009). These theories are obviously less interesting given the discovery of a Higgs-like boson, but it is extraordinarily difficult, and perhaps impossible, to resolve whether the Higgs boson is a fundamental scalar or merely a composite particle acting like a scalar. Also, theories with no true Higgs boson can have another particle—a dilation, for example—that acts like a Higgs boson.

Therefore, these theories still have life within them, and more data is required to gain confidence in these alternative explanations or rule them out.

Nevertheless, the least complicated thoughts suggest to us that a simple Higgs boson has been discovered with mass of approximately 126 GeV (Aad et al. 2012; Chatrchyan et al. 2012). Of course, there is no certainty that it is the SM Higgs boson. Indeed, such certainty is likely to never exist, but measurements at the LHC can likely give us confidence that its couplings are within 20 % of the values that the SM Higgs boson would have. Next-generation colliders, such as an e^+e^- linear collider, would be able to further refine this to percent level, or perhaps even show that there are small deviations from SM expectations. In any case, it is legitimate to call it "a Higgs boson" since it appears to be coupling to the vector boson and fermions according to their mass values, and that puts an added confidence that the particle is associated with mass generation. Again, metaphysical certainty into the nature of any particle will always be out of the question, but the evidence is accruing and the words "for all practical purposes" are just around the corner.

This has been a major achievement by humans. The historical theory development that culminated in a highly speculative prediction for a new Higgs boson that turned out to be there is just one aspect of this achievement. There is also the decades of work and expertise built up to invent and apply experimental techniques that discovered the boson. This is not to mention the impressive human resource management skills needed to herd all the people together in a collective effort to divide tasks and construct the coherent whole—the discovery.

The smugness we may feel for the discovery of the Higgs boson is to be tempered with the stark truth that nothing else has been found at the LHC at this time. If it continues this way it means that many predictions, influenced by concepts of effective theories, were wrong that insisted that the Higgs boson needed an entourage of other particles very close by in mass to tame its quantum instabilities. Maybe they were only wrong quantitatively, and new particles and dynamics are around the corner to vindicate effective theories.

Or perhaps there is yet another factor that is overriding our effective theory intuitions. Perhaps there is a multiverse where the solution to the Hierarchy Problem suggests that large statistics of finetuned solutions dominate over the fewer number of non-tuned solutions in the landscape, leading to a higher probability of our Universe landing in a highly tuned solution ($c \ll 1$). Thus, guided by concerns over the cosmological constant problem, it has been suggested that this statistical, stringy naturalness over the landscape may take precedence over normal naturalness envisioned from effective field theories (Douglas 2007; Kumar 2006). Although not directly related to external particle physics interactions, the cosmological constant can be considered as the coefficient of yet another gauge-invariant, Lorentz-invariant operator—the operator being merely a constant: $-\mathscr{L}_{cc} = \Lambda_{cc}^4$. The tiny value of this coefficient, $\Lambda_{cc}^4 \simeq (10^{-3} \text{ eV})^4$, is well below any conceivable theory expectation. It is the elephant in the room for effective field theories. However, it is an unexpressed article of faith among most particle physicists that the solution to the Cosmological Constant Problem lies in the details of mysterious quantum gravity, and that the new concepts buried in that unknown solution do not materially affect the natural solution to the Hierarchy Problem. Landscapists question that assumption. This is controversial with conflicting claims over unrealistic theories; nevertheless, it is an interesting idea that might one day be impactful.

Data keeps coming, and searches for new particles that would vindicate our most basic notions of effective field and naturalness continue. Many "good ideas" are now dead after years of data have found no evidences for them. There is no theorem that we will have full resolution to all the "good ideas" within our lifetimes, or that any of the colliders we are running or contemplating in the future will have enough energy or luminosity or precision to give a final say on the matter. Nevertheless, the field carries on and the tree of various interpretations of what has been seen and what has not been seen grows branches, flowers, and surely will bear fruit again.

References

Aad, G., et al.: [ATLAS Collaboration] (2012), arXiv:1207.7214
Altarelli, G.: arXiv:0711.0161 (2007)
Arkani-Hamed, N., Dimopoulos, S.: JHEP **0506**, 073 (2005) [arXiv:hep-th/0405159]
Arkani-Hamed, N., Dimopoulos, S., Giudice, G.F., Romanino, A.: Nucl. Phys. B **709**, 3 (2005) [arXiv:hep-ph/0409232]
Chatrchyan, S., et al.: [CMS Collaboration] (2012), arXiv:1207.7235
Chivukula, R.S., Dobrescu, B.A., Georgi, H., Hill, C.T.: Phys. Rev. D **59**, 075003 (1999) [arXiv:hep-ph/9809470]
Clauset, A.: How large should whales be?. arxiv:1207.1478 (2012) (Interesting environmental effects nevertheless may be at play. See, for example, the speculations)
Csaki, C., Grojean, C., Pilo, L., Terning, J.: Phys. Rev. Lett. **92**, 101802 (2004a) [arXiv:hep-ph/0308038]
Csaki, C., Grojean, C., Murayama, H., Pilo, L., Terning, J.: Phys. Rev. D **69**, 055006 (2004b) [arXiv:hep-ph/0305237]
Cui, Y., Gherghetta, T., Wells, J.D.: arXiv:0907.0906 [hep-ph] (2009)
De Gouvea, A.: hep-ph/0411274 (2004)
Donoghue, J.F., Golowich, E., Holstein, B.R.: Dynamics of the Standard Model. Cambridge University Press, Cambridge (1992). (One example of an excellent pedagogical introduction to the chiral lagrangian effective theory can be found in chapter IV)
Douglas, M.R., Kachru, S.: Rev. Mod. Phys. **79**, 733 (2007) (sec. II.F.3) [arXiv:hep-th/0610102]
Frampton, P.H., Vafa, C.: arXiv:hep-th/9903226 (1999) (For an exploratory vision of possibilities)
Giudice, G.F., Romanino, A.: Nucl. Phys. B **699**, 65 (2004) [Erratum-ibid. B 706, 65 (2005)] [arXiv:hep-ph/0406088]
Giudice, G.F.: arXiv:0801.2562 [hep-ph] (2004)
Goldstone, J., Salam, A., Weinberg, S.: Phys. Rev. **127**, 965 (1962)
Grossman, Y.: hep-ph/0305245 (2003) (For dedicated neutrino physics reviews)
Hill, C.T.: Phys. Lett. B **266**, 419 (1991)
Hill, C.T.: Phys. Lett. B **345**, 483 (1995) [arXiv:hep-ph/9411426]
Kayser, B: Neutrino mass, mixing, and flavor change. In: Nakamura, K., et al. [Particle Data Group Collaboration], Review of Particle Physics. J. Phys. G G **37**, 075021 (2012) (For a summary of neutrino masses and mixing constraints)
Kumar, J.: Int. J. Mod. Phys. A **21**, 3441 (2006) (sec. 3.4) [arXiv:hep-th/0601053]
Lane, K., Martin, A.: arXiv:0907.3737 [hep-ph] (2009) (For a recent approach to technicolor)
Martin, S.P.: A Supersymmetry Primer. hep-ph/9709356 (1997)

Martin, S.P.: Phys. Rev. D **46**, 2197 (1992) [arXiv:hep-ph/9204204]

Mohapatra, R.N.: hep-ph/0412050 (2004)

Peskin, M.E., Schroeder, D.V.: An Introduction to Quantum Field Theory. Persueus Books, Reading (1995) (A recent excellent description)

Polchinski, J.: arXiv:hep-th/9210046 (1992)

Rattazzi, R.: arXiv:hep-ph/0607055 (2006)

Renton, P.: Electroweak Interactions. Cambridge University Press, Cambridge (1990). (For a brief technical review of this, see chapter 5)

Schmaltz, M., Tucker-Smith, D.: Ann. Rev. Nucl. Part. Sci. **55**, 229 (2005) [arXiv:hep-ph/0502182]

Sundrum, R.: arXiv:hep-th/0508134 (2005) (For theory reviews of extra dimensions)

Wells, J.D.: arXiv:hep-ph/0306127 (2003)

Wells, J.D.: Phys. Rev. D **71**, 015013 (2005) [arXiv:hep-ph/0411041]

Wilson, K.G., Kogut, J.B.: Phys. Rep. **12**, 75 (1974) (For an early exposition from Wilson)

Chapter 5
Effective Theories and Theory Choice

Abstract Promoting a theory with a finite number of terms into an effective field theory with an infinite number of terms worsens simplicity, predictability, falsifiability, and other attributes often favored in theory choice. However, the importance of these attributes pales in comparison with consistency, both observational and mathematical consistency, which propels the effective theory to be superior to its simpler truncated version of finite terms, whether that theory be renormalizable (e.g., Standard Model of particle physics) or unrenormalizable (e.g., gravity). Some implications for the Large Hadron Collider and beyond are discussed, including comments on how directly acknowledging the preeminence of consistency can affect future theory work.

5.1 Introduction

One of the most interesting questions in philosophy of science is how to determine the quality of a theory. Given the data, how can we infer a "best explanation" for the data. This often goes by the name "Inference to Best Explanation" (IBE) (Harman 1965; Lipton 1991; Clayton 1997). The wide variety of claims for important criteria are a measure of how difficult it is to come up with a clear and general algorithm for choosing between theories. Some claim even that it is intrinsically not possible to come up with a methodology of deciding (Lehrer 1974; Newton-Smith 1981). Nevertheless the goals of IBE are worthy, and the payoff is high upon increased understanding, if for no other reason than the extraphilosophical importance of distributing grant money more fairly to researchers. Furthermore, whether objective criteria for IBE are possible, all practitioners of science have no choice but to engage in the "infer" part even if they may never touch upon the "best explanation" part of IBE.

The goal of this chapter is to survey theory choice criteria in the context of effective theories. It has been accepted by the physics communities that theories must be

---

J. D. Wells, *Effective Theories in Physics*, SpringerBriefs in Physics,
DOI: 10.1007/978-3-642-34892-1_5, © The Author(s) 2012, corrected publication 2022

"effectified", that is they must be augmented to include all possible interactions consistent with the stated symmetries to all orders. On the surface the resulting effective theories are in conflict with the rules of IBE, whether they be the murky rules that some physicists put forward when they talk about theory choice, or the precisely stated rules developed by philosophers. Upon closer inspection effective theories rise quickly to the top in theory choice when admitting to the primacy of *consistency* in theory choice. That is the claim, to be developed below. The reader should be warned that I will use the acronym IBE to mean any attitude, theory, system by which people decide that one theory is a better description of nature than another, or that a theory under consideration is a good theory at all.

5.2 The Standard Model's Triumphs and Woes

This chapter is primarily written from the science perspective of elementary particle theory, with particular emphasis on the subfield "beyond the Standard Model physics". In this subfield, the task is to look out over nature and ask what is not adequately described by the Standard Model (SM) of particle physics. The Standard Model has been with us for about 40 years. It consists of three families of up-type quarks (u, c, t), three families of down-type quarks (d, s, b), three families of leptons (e, μ, τ) and three families of neutrinos (ν_e, ν_μ, ν_τ). These interact with each other according to gauge field theory interactions, mediated by the force carrier bosons of the photon, gluons and W and Z bosons. Every particle that has mass is said to achieve it by a condensing Higgs boson. For a more complete non-technical or technical description of the SM see references Kane (1996) and Griffiths (2008), respectively.

The SM is a renormalizable theory and can be fully described on one page using standard nomenclature of mathematics and relativistic quantum field theory. Despite that simplicity, it can account for every measurement ever made at high-energy colliders. It is an enormous human achievement. So why are there researchers searching for theories "beyond the SM"? There are many reasons, of which I will name a few:

- There are non-collider observations we still cannot explain such as galactic rotation measurements that imply the existence of dark matter, and the preponderance of baryons over anti-baryons in the universe.
- The particle content and the three gauge forces cry out for unification (e.g., grand unified theories).
- There are many of parameters with large hierarchies that beg for explanation ($m_t/m_e > 10^6$).
- The SM Higgs boson appears unstable to quantum corrections and is thus unnatural.
- Surely there is more than just the SM (e.g., SM is just copies of stuff in our bodies).
- Embedding gravity into quantum mechanics is a severe challenge and should bring new implications to the particle physics world (e.g., string theory).

Thus, there are many opportunities to devise new theories that solve one or more of these problems. The theories are necessarily speculative upon birth. They are put to the test, and the simple fact is that at any given moment there are a multitude of theories that appear to be able to solve one or more of the issues. We have many variants of supersymmetric theories, strongly coupled theories, extra dimensional theories, etc. that appear to be able to do the job and are not yet distinguishable by currently known data. How does a scientist determine which is the best? The rules are not clear, of course, and we shall first ask how do scientists make theory choices, and when does IBE enter their calculus.

5.3 Theory Choice Among Practitioners

Typically a particle physicist will look at the SM problems listed above and set out to construct a new theory that explains one or more of them. The particle researcher often stumbles into a theory choice of what to work on not based on IBE but rather DBO (deduction of best opportunities). The opportunities that arise may include matching yourself with the best PhD advisor who is working on theory X, researching a fashionable topic to get a good job, supporting a clever theory that the researcher devised that might not have high probability of being correct but has highest probability of enormous personal pay-off, etc. The last reason then circles back on the first reason as advisors ask their students to work on theories that they themselves devised. Furthermore, the subtleties of elementary particle physics and beyond the SM theories are such that it could take new practitioners years before they feel confident that they could make a reliable IBE estimate even if the criteria for such were clear to them. Thus, IBE considerations are often not the dominant force for their theory choice (i.e., what to work on) in a practicing scientist's career.

IBE issues do arise when there is competition among researchers for journal space, research funds, and conference time slots. IBE-like arguments ensue. Words used to describe the evaluation of theories are familiar to philosophers: simplicity, economy, calculability, compatibility with data, testability, falsifiability, naturalness, finetuning, predictivity, unification, no ad hoc assumptions, etc. Researchers become attorneys for their theories and weight the various IBE criteria which most favorably supports the direction of their research lines. That is why experimentalists and phenomenologists emphasize "falsifiability" and "observational consistency" much more than string theorist, who emphasize "unification", "completeness" and "mathematical consistency".

It is often said at the end of arguments between theorists about their pet theories that "experiment will decide". However, as experiments become larger and costs grow steeply, the time frame may extend well past decades to even centuries. It took more than 25 years for CERN to conceive and build the LHC, for example. There is no guarantee that any timeline convenient to a human is relevant to future experimental construction. However, what is relevant to human time scales is deciding what are "better" or "best theories", since we should use that to allocate resources of time, money, etc. Working toward perfecting IBE criteria, no matter how controversial they are, is clearly warranted.

The recent universal acclaim of effective theories gives us an opportunity to apply IBE thinking to a case that is not controversial as a means to better understand the weight that should be given to various elements of IBE. In the next section, I will describe how the effective SM is different from the SM, and then we shall survey their IBE qualities, with an eye toward gaining insight along the way.

5.4 The Standard Model Versus the Effective Standard Model

The SM has a finite number of operators of dimension four or less. The effective SM (ESM) is the SM but with all possible higher dimensional operators present consistent with the sacrosanct symmetries of the SM: Lorentz symmetry and $SU(3)_c \times SU(2)_L \times U(1)_Y$ gauge symmetries. Thus we can relate the lagrangians of the two by the equation

$$\mathscr{L}_{ESM} = \mathscr{L}_{SM} + \sum_{n,i} \eta_{n,i} \frac{\mathscr{O}_i^{(4+n)}}{\Lambda^n}, \qquad (5.1)$$

where $\mathscr{O}_i^{(4+n)}$ is the collection of all operators of higher dimension $4+n$ that respect the symmetries of the SM and have unknown couplings $\eta_{n,i}/\Lambda^n$ in front.

The SM matches all observed high-energy collider data to excellent compatibility. There can be no additional operators that would improve the fit by a meaningful amount. Furthermore, if any of the higher-dimensional operators of $\mathscr{L}_{ESM}$ become worrisome with respect to the data, we need merely tune down the strength of the interaction by making its associated $\eta_{n,i}$ coupling smaller, to escape the problem.

Which theory is better, the ESM or SM, given that they both can be made equally compatible with the data? To answer this question, let us first apply some of the IBE thinking common in the particle physics community. Our example source for a typical particle physicist approach to these issues will be the essay written by Nobel Laureate Burton Richter (2006). We shall also attempt to answer the question using the criteria of the philosopher Paul Thagard (1978), whose paper is still considered one of the key early expositions on theory choice criteria for IBE.

5.5 Richter's IBE Criteria

There are not many official forums through which practicing particle physicists are encouraged to divulge their IBE criteria. But one forum where it regularly happens, both in essays and in letters to the editor, is in professional society monthly notices.

One of the most talked about articles of this kind in recent years was written by Burton Richter (2006). Richter and Sam Ting won the Nobel prize of physics in 1976 for finding the J/ψ particle, which was a key discovery in establishing the SM.

Richter has been horrified by what he views are "major problems in the philosophy behind theory" research. He says,

> Simply put, most of what currently passes as the most advanced theory looks to be more theological speculation, the development of models with no testable consequences, than it is the development of practical knowledge, the development of models with testable and falsifiable consequences (Karl Popper's definition of science).

Richter goes on to say that more weight should be put on

1. theories that have testable and *falsifiable* consequences, and
2. theories that *simplify* rather than increase complication.

Incidentally, he also discusses two anti-criteria that should not be used, which are the anthropic principle and naturalness. Let us not discuss these anti-criteria, but rather judge the SM versus ESM based on what Richter would have us do, on falsifiability and simplicity.

Regarding falsifiability and testable consequences, an argument can be made that the SM wins. The ESM has an infinite number of operators with coefficients to be pinned down by data later, and as such can accommodate more experimental outcomes the SM. After all, the ESM reduces to SM when $\Lambda^n \to \infty$. Thus, the SM is much more testable and falsifiable than ESM.

Although not central to the subsequent discussion, I would like to remark that *falsifiability* has never struck me as strong argument for theory deciding for two reasons. Skepticism toward falsifiability has long been held in the philosophy community, but let me give what I think are two strong reasons to worry about its applicability. Let's take an example of an unfalsifiable theory: Theory X says that obvious fact Y is true (e.g., emeralds are green, or something trivially true like that), *and* that angels live in another universe. We can use this silly theory to illustrate why falsifiability is not a very solid criteria.

First, the modularity of the theory can be under dispute, such as the more testable first statement versus the second statement. Second, if things change dramatically such that what was true yesterday is not true tomorrow (tomorrow Y is false), then the theory is trivially invalidated. Does that make it falsifiable? In that case, all theories are falsifiable by scattering the word "always" through-out its description. And third, it is never clear if *falsifiability* must be applicable in principle or in practice. In principle perhaps everything is *falsifiable* (e.g., many versions of string theory—just run a collider at 10^{19} GeV), whereas in practice good theories might not be (perhaps: string theory, high scale warped extra dimensions, etc.) because of lack of money, technology, time, or manpower to test it.

In short, you can like it, you can hope for it, you can wish for it, you can say it would make our lives as scientists much easier if so, but it would presumptuous of us to say that Nature cares one whit if we can falsify a true statement. Nevertheless, we shall take it seriously because we are investigating somebody else's criteria, which happen to be shared by many others. And as we have noted above, the testable and falsifiability criteria favors SM over ESM.

Regarding *simplicity*, the SM has a finite number of terms with a finite number of coefficients, whereas the ESM has an infinite number of terms with an infinite number of coefficients. No contest, SM wins in the simplicity category.

According to Richter's two key criteria, falsifiability and simplicity, the SM is the winner, and we infer it to be the best theory.

5.6 Thagard's IBE Criteria

The philosophy literature is vast on the subject. Surveying it with sweeping scope would not be enlightening and picking just one approach to compare leaves one wanting. Nevertheless, I will do the latter, choosing a classic paper on the subject by Paul Thagard (1978).

Thagard's theory choice criteria are

1. *Consilience*: The measure of how many facts the theory explains; furthermore, "a consilient theory unifies and systematizes."
2. *Simplicity*: The quality of having the fewest "auxiliary hypothesis," fewest ad hoc additions, and most ontological economy.
3. *Analogy*: Shared characteristics between two theories, leads to one theory admitting a new characteristic if the new characteristic is part of the other theory and explains the shared characteristics there.

Regarding *consilience*, it is a draw between SM and ESM. The facts are equally compatible in the two theories, and there is no relevant advantage in either in the realm of unification and systematizing.

Regarding *simplicity*, although it is not exactly the kind of simplicity that Richter was talking about, the SM clearly is superior to the ESM in this category. The new operators of the ESM simply add more.

Regarding *analogy*, it is my view that the ESM wins out over the SM. I will explain why twice. First, I will explain it here strictly in the language of Thagard's analogy propositions. I will explain it a second time later heuristically using particle physics language from Steven Weinberg.

As Thagard explains, by *analogy* he does not mean the standard syllogism

A is P, Q, R, S
B is P, Q, R
Thus, B is S.

No, he means something more causally connected:

A is P, Q, R, S
B is P, Q, R
If S explains P, Q, R in A, then B is S.

I will follow this analogy criteria by first defining

A = chiral lagrangian of pion scattering

B = Effective Standard Model (ESM).

The chiral lagrangian of pion scattering is

$$\mathcal{L}_\pi = \frac{v^2}{4} Tr\left(\partial_\mu U \partial^\mu U^\dagger\right) + \frac{v^2}{\Lambda^2}\left[Tr\left(\partial_\mu U \partial^\mu U^\dagger\right)\right]^2 + \cdots \qquad (5.2)$$

where

$$U \equiv \exp(i\tau \cdot \pi/v^2). \qquad (5.3)$$

This lagrangian has an infinite number of terms respecting the underlying $SU(2)$ custodial symmetries. The pions are the π fields in U, and v is the vacuum expectation breaking of the custodial symmetry $SU(2)_L \times SU(2)_R$ to its $SU(2)_V$ vector subgroup. All the interactions of the pions are contained within these terms. As the energy increases the higher order terms in the lagrangian become more important, and the data can be accommodated. This theory was very useful. It was determined that all the higher order corrections needed to be there, although a deep appreciation of why was not to come until Ken Wilson's renormalization breakthroughs years later.

Now, the shared properties P, Q, R of the chiral lagrangian theory of pion scattering and the ESM are

P, Q, R = quantum field theory, perturbative expansion theory, all lowest dimensionality terms allowed by symmetries of the theory are present, finite number of terms relevant in deep infrared, etc.

and the new characteristic S in theory A is

S = all operators consistent with the symmetries are present.

S explains P, Q, R because relevant terms are a subset of "all operators".

Thus, by Thagard's *analogy* we would say that the SM should be augmented by all possible terms consistent with its symmetries $\longrightarrow$ ESM. The argument is further strengthened later when we catch Weinberg directly using the language of analogy to support the generalization of effective field theory techniques to the SM.

The result of our analysis based on Thagard's IBE criteria is SM +1, ESM +1, and Draw +1. No clear resolution to be found here.

5.7 Non-negotiable Attributes of a Best Explanation

What is lacking in our discussion of IBE criteria is a rank ordering of attributes. We must first ask ourselves what is non-negotiable. *Falsifiability* is clearly something that can be haggled over. *Simplicity* is subject to definitional uncertainty, and furthermore has no universally accepted claim to preeminence. *Naturalness, calculability, unifying ability, predictivity*, etc. are also subject to preeminence doubts.

What is non-negotiable is *consistency*. A theory shown definitively to be inconsistent does not live another day. It might have its utility, such as Newton's theory of gravity for crude approximate calculations, but nobody would ever say it is a better theory than Einstein's theory of General Relativity.[1]

Consistency has two key parts to it. The first is that what can and has been computed must be consistent with all known observational facts. As Murray Gell-Mann said about his early graduate student years, "Suddenly, I understood the main function of the theoretician: not to impress the professors in the front row but to agree with observation (Gell-Mann 1994)." Experimentalists of course would not disagree with this non-negotiable requirement of *observational consistency*. If you cannot match the data what are you doing, they would say?

However, theorists have a more nuanced approach to establishing *observational consistency*. They often do not spend the time to investigate all the consequences of their theories. Others do not want to "mop up" someone else's theory, so they are not going to investigate it either. We often get into a situation of a new theory being proposed that solves one problem, but looks like it might create dozens of other incompatibilities with the data but nobody wants to be bothered to compute it. Furthermore, the implications might be extremely difficult to compute.

Sometimes there must be suspended judgment in the competition between excellent theories and observational consequences. Lord Kelvin claimed Darwin's evolution ideas could not be right because the sun could not burn long enough to enable long-term evolution over millions of years that Darwin knew was needed. Darwin rightly ignored such arguments, deciding to stay on the side of geologists who said the earth appeared to be millions of years old (Gavin et al. 2008). Of course we know now that Kelvin made a bad inference because he did not know about the fusion source of burning within the sun that could sustain its heat output for billions of years.

A second part to *consistency* is *mathematical consistency*. There are numerous examples in the literature of subtle mathematical consistency issues that need to be understood in a theory. Massive gauge theories looked inconsistent for years until the Higgs mechanism was understood. Some gauge theories you can dream up are "anomalous" and inconsistent. Some forms of string theory are inconsistent unless there are extra spatial dimensions. Extra time dimensions appear to violate causality, even when one tries to demand it from the outset, thereby rendering the theory inconsistent. Theories with ghosts, which may not be obvious upon first inspection, give negative probabilities of scattering.

Mathematical consistency is subtle and hard at times, and like *observational consistency* there is no theorem that says that it can be established to comfortable

---

[1] The word "better" in this context can induce apoplectic shocks in pedants. To avoid that, by "better" I wish to say that it is closer to the true, underlying theory, whatever that may mean or be. I do not wish it to mean "better to calculate a hammer fall on the moon in under three lines for primary school children", or any other similar appeal to convenience or simplicity.

levels by theorists on time scales convenient to humans. Sometimes the inconsistency is too subtle for the scientists to see right off. Other times the calculability of the mathematical consistency question is too difficult to give definitive answer and it is a "coin flip" whether the theory is ultimately consistent or not. For example, pseudomoduli potentials that could cause a runaway problem are incalculable in some interesting dynamically broken supersymmetric theories (Intriligator et al. 2009).

It is not controversial that *observational consistency* and *mathematical consistency* are non-negotiable; however, the due diligence given to them in theory choice is often lacking. The establishment of *observational consistency* or *mathematical consistency* can remain in an embryonic state for years while research dollars flow and other IBE criteria become more motivational factors in research and inquiry, and the *consistency* issues become taken for granted.

This is one of the themes of Gerard 't Hooft's essay "Can there be physicist without experiments?" (Hooft 2001). He reminds the reader that some of the grandest theories are investigations of the nature of spacetime at the Planck scale, which is many orders of magnitude beyond where we currently have direct experimental probes. If this is to continue as a physics enterprise it "may imply that we should insist on much higher demands of logical and mathematical rigour than before." Despite the weakness of verb tense employed, it is an incontestable point. It is in these Planckian theories, such as string theory and loop quantum gravity, where the lack of *consistency* rigor is so plainly unacceptable. However, the cancer of lax attention to consistency can spread fast in an environment where theories and theorists are fêted before vetted.

5.8 Effective Field Theories and Consistency

Let us begin with the claim at the heart of our discussion. The claim behind the ascendancy of effective theories is that unless there is good and explicit reason otherwise, consistency requires that a theory have all possible interactions consistent with its symmetries at every order.

The claim has its origins in the work of Wilson, whose original review article with Kogut (Wilson and Kogut 1974) is a classic. There are many modern reviews of effective theories that make or assume the above claim (Polchinski 1992; Cohen 1993; Rothstein 2003). Weinberg's recent historical perspective (Weinberg 1964) gives an excellent summary of what was learned:

> I was struck [at Erice school in 1976] by Kenneth Wilson's device of "integrating out" short-distance degrees of freedom by introducing a variable ultraviolet cutoff, with the bare couplings given a cutoff dependence that guaranteed that physical quantities are cutoff independent. Even if the underlying theory is renormalizable, once a finite cutoff is introduced it becomes necessary to introduce every possible interaction, renormalizable or not, to keep physics strictly cutoff independent.... Indeed, I realized that even without a cutoff, as long as every term allowed by symmetries is included in the Lagrangian, there will always be counterterm available to absorb every possible ultraviolet divergence....

Therefore, consistency of the theory—the absorption of ultraviolet divergences, the maintaining of independence of arbitrary ultraviolet scale cutoff, etc.—requires the introduction of all possible terms allowed by the symmetries.

The issue of consistency then becomes front and center, and the issues of simplicity and testability fade in importance. From our discussion above we know that without this important issue of consistency, the effective SM may not win in a theory choice competition compared to the SM with just its renormalizable operators, since it worsens the otherwise positive features of simplicity and testability. Therefore, the establishment of rigorous consistency requirements on the theory were crucial in the decision.

5.9 Relation to Thagard's Analogy Criterion

I would like to take a quick aside and show that physicists do reason in real-life, complex theory circumstances through the *analogy* criterion of Thagard. Indeed, it is a separate argument for the general applicability of effective theories.

In the same historical review article (Weinberg 1964) quoted above, Weinberg shows that because effective field theory ideas were necessary in chiral dynamics (low-energy pion scattering), the concept should also apply to the SM. Here is a relevant quote:

> Perhaps the most important lesson from chiral dynamics was that we should keep an open mind about renormalizability. The renormalizable Standard Model of elementary particles may itself be just the first term in an effective field theory that contains every possible interaction allowed by Lorentz invariance and the $SU(3) \times SU(2) \times U(1)$ gauge symmetry, only with the non-renormalizable terms suppressed by negative powers of some very large mass M, just as the terms in chiral dynamics with more derivatives ... are suppressed by negative powers of $2\pi F_\pi \simeq m_N$.

One should note the usage of analogy language: "most important lesson from chiral dynamics" and "just as in the terms in chiral dynamics". Thus, the syllogistic representations given in Sect. 5.6 are shown to apply and be part of theory choice for particle physicists.

5.10 Summary: The Preeminence of Consistency

I will conclude by stating my two central points that generalize the discussion we have had above in comparing the effective SM with the SM.

My first point is that the conditions of theory choice should be ordered. Frequently we see the listing of criteria for theory choice given in a flat manner, where one is not given precedence over the other a priori. We see consilience, simplicity, falsifiability, naturalness, consistency, economy, all together in an unordered list of factors when judging a theory. However, *consistency* must take precedence over any other factors. Observational consistency is obviously central to everyone, most especially our experimental colleagues, when judging the relevance of theory for describing nature. Despite some subtleties that can be present with regards to observational consistency[2] it is a criterion that all would say is at the top of the list.

Mathematical consistency, on the other hand, is not as fully appreciated. In Richter's essay excoriating theorists he did not appear to recognize or acknowledge the central role that mathematical consistency plays in developing and vetting theories. Mathematical consistency has a preeminent role right up there with observational consistency, and can be just as subtle, time-consuming and difficult to establish. We have seen that in the case of effective theories it trumps other theory choice considerations such as simpleness, predictivity, testability, etc.

My second point builds on the first. Since consistency is preeminent, it must have highest priority of establishment compared to other conditions. Deep, thoughtful reflection and work to establish the underlying self-consistency of a theory takes precedence over finding ways to make it more natural or to have less parameters (i.e., simple). Highest priority must equally go into understanding all of its observational implications. A theory should not be able to get away with being fuzzy on either of these two counts, before the higher order issues of simplicity and naturalness and economy take center stage. That this effort might take considerable time and effort should not be correlated with a theory's value, just as it is not a theory's fault if it takes humans decades to build a collider to sufficiently high energy and luminosity to test it.

Additionally, dedicated effort on mathematical consistency of the theory, or class of theories, can have enormous payoffs in helping us understand and interpret the implications of various theory proposals and data in broad terms. An excellent example of that in recent years is by Adams et al. (2006), who showed that some theories in the infrared with a cutoff cannot be self-consistently embedded in an ultraviolet complete theory without violating standard assumptions regarding superluminality or causality.

5.11 Implications for the LHC and Beyond

Finally, I would like to make a comment about the implications of this discussion for the LHC and other colliders that may come in the future. First, it is obvious that we must be prepared for and search for higher-dimensional operators in the effective SM that goes beyond the relevant and marginal operators of the SM. This is indeed happening at the LHC, and first indications of new physics may very well come from

---

[2] There can be circumstances where a theory is observationally consistent in a vast number of observables, but in a few it does not get right, yet no other decent theory is around to replace it. In other words, observational consistency is still the top criterion, but the best theory may not be 100 % consistent.

small perturbations in SM observables due to the subtle effects of these suppressed operators.

However, there is broader point to be made regarding implications for colliders. In the years since the charm quark was discovered in the mid 1970s there has been tremendous progress experimentally and important new discoveries, including the recent discovery of a Higgs boson-like state (Aad et al. 2012; Chatrchyan et al. 2012), but no dramatic new discovery that can put us on a straight and narrow path beyond the SM. That may change soon at the LHC. Nevertheless, it is expensive in time and money to build higher energy colliders, our main reliable transporter into the high energy frontier. This limits the prospects for fast experimental progress.

In the meantime though, hundreds of theories have been born and have died. Some have died due to incompatibility of new data (e.g., simplistic technicolor theories, or simpleminded no-scale supersymmetry theories), but others have died under their own self-consistency problems (e.g., some extra-dimensional models, some string phenomenology models, etc.). In both cases, it was care in establishing consistency with past data and mathematical rigor that have doomed them. In that sense, progress is made. Models come to the fore and fall under the spotlight or survive. When attempting to really explain everything, the consistency issues are stretched to the maximum. For example, it is not fully appreciated in the supersymmetry community that it may even be difficult to find a "natural" supersymmetric model that has a high enough reheat temperature to enable baryogenesis without causing problems elsewhere (Olechowski et al. 2009; Covi et al. 2011). There are many examples of ideas falling apart when they are pushed very hard to stand up to the full body of evidence of what we already know.

Relatively speaking, theoretical research is inexpensive. It is natural that a shift develop in fundamental science. The code of values in theoretical research will likely alter in time, as experimental input slows. Ideas will be pursued more rigorously and analysed critically. Great ideas will always be welcome. However, soft model building tweaks for simplicity and naturalness will become less valuable than rigorous tests of mathematical consistency. Distant future experimental implications identified for theories not fully vetted will become less valuable than rigorous computations of observational consistency across the board of all currently known data. One can hope that unsparing devotion to full consistency, both observational and mathematical, will be the hallmarks of the future era.

References

Aad, G., et al. [ATLAS Collaboration]: Phys. Lett. B **716**, 1 (2012) [arXiv:1207.7214 [hep-ex]]
Adams, A., Arkani-Hamed, N., Dubovsky, S., Nicolis, A., Rattazzi, R.: JHEP **0610**, 014 (2006) [hep-th/0602178]
Chatrchyan, S., et al. [CMS Collaboration]: Phys. Lett. B **716**, 30 (2012) [arXiv:1207.7235 [hep-ex]]
Clayton, P.: Inference to the best explanation. Zygon **32**, 377 (1997)
Cohen, A.G.: Selected topics in effective field theories for particle physics. In: Proceedings of the Theoretical Advanced Study Institute (TASI 1993), Boulder, Colorado (1993)

Covi, L., et al.: JHEP **1101**, 033 (2011) [arXiv:1009.3801]

Gavin, S., Conn, J., Karrer, S.P.: The Age of the Sun: Kelvin vs. Darwin. Wayne State University (2008)

Gell-Mann, M.: The Quark and the Jaguar. Little, Brown and Company, New York (1994)

Griffiths, D.: Introduction to Elementary Particles, 2nd edn. Wiley, New York (2008)

Harman, G.H.: Inference to the best explanation. Philos. Rev. **74**, 88 (1965)

Intriligator, K., Shih, D., Sudano, M.: Surveying pseudomoduli: the good, the bad and the incalculable. JHEP **0903**, 106 (2009) [arXiv:0809.3981 [hep-th]]

Kane, G.: The Particle Garden. Basic Books, New York (1996)

Lehrer, K.: Knowledge. Clarendon Press, Oxford (1974)

Lipton, P.: Inference to Best Explanation. Oxford University Press, Oxford (1991)

Newton-Smith, W.H.: The Rationality of Science. Routledge, London (1981)

Olechowski, M., et al.: JHEP **0912**, 026 (2009) [arXiv:0908.2502]

Polchinski, J.: Effective field theory and the fermi surface [arXiv:hep-th/9210046]

Richter, B.: Theory in particle physics: theological speculation versus practical knowledge. Phys. Today **59**, 8 (2006)

Rothstein, I.Z.: TASI lectures on effective field theories. In: Proceedings of the Theoretical Advanced Study Institute (TASI 2003), Boulder, Colorado (2003)

Hooft, G.'t: Int. J. Mod. Phys. A **16**, 2895 (2001)

Thagard, P.R.: The best explanation: criteria for theory choice. J. Philos. **75**, 76 (1978)

Weinberg, S.: Effective field theory, past and future [arXiv:0908.1964 [hep-th]]

Wilson, K.G., Kogut, J.B.: Phys. Rep. **12**, 75 (1974)

Correction to: Effective Theories in Physics

Correction to:
**J. D. Wells, *Effective Theories in Physics*, SpringerBriefs
in Physics, https://doi.org/10.1007/978-3-642-34892-1**

The updated version of the book can be found at
https://doi.org/10.1007/978-3-642-34892-1

© The Author(s) 2022
J. D. Wells, *Effective Theories in Physics*, SpringerBriefs in Physics,
https://doi.org/10.1007/978-3-642-34892-1_6

Printed in the United States
by Baker & Taylor Publisher Services